IPT's HAN
TRAININ(

D1536839

For Pricing And Ordering Information Contact:

IPT PUBLISHING AND TRAINING LTD.
BOX 9590, EDMONTON, ALBERTA CANADA T6E 5X2
www.iptbooks.com
Email: iptpub@interbaun.com

Phone (780) 962-4548, Fax (780) 962-4819
Toll Free (888) 808-6763

Accepted Ordering and Payment Methods:
- Personal Cheques
- Money Orders
- Company Purchase Orders
- Visa
- Mastercard

BOOKS PUBLISHED BY IPT

HANDBOOK and TRAINING MANUAL
VERSIONS AVAILABLE FOR EACH

METAL TRADES

INDUSTRIAL TRADES

INDUSTRIAL FASTENERS

CRANE AND RIGGING

PIPE TRADES

ELECTRICAL

ROTATING EQUIPMENT

SAFETY FIRST

BASIC HYDRAULICS

BLUEPRINT INTERPRETATION
(One format only)

IPT's
ELECTRICAL HANDBOOK

by

HERB PUTZ (B.Ed., RET, Master Electrician)

Published by
**IPT PUBLISHING AND TRAINING LTD.
BOX 9590, EDMONTON, ALBERTA CANADA T6E 5X2
www.iptbooks.com
Email: iptpub@compusmart.ab.ca**

**Phone (780) 962-4548, Fax (780) 962-4819
Toll Free 1-888-808-6763**

Printed by
Quebecor Jasper Printing, Edmonton, Alberta, Canada

The material presented in this publication has been prepared in accordance with recognized trade working practices and codes, and is for general information only. In areas of critical importance the user should secure competent engineering advice with respect to the suitability of the material contained herein, and comply with the various codes, standards, regulations or other pertinent legal obligation. Anyone utilizing this information assumes all responsibility and liability arising from such use.

First Printing, May 1994

Second Printing, March 1997

Third Printing, January 2003

ISBN #0-920855-22-9

COPYRIGHT © 1994

Metric Content and Electrical Codes

The author faced a difficult task in determining the degree to which the Metric system should be applied in this publication. This book is primarily designed for Canada and the United States. Both countries have adopted and use the Metric system, however the degree of use is not consistent in all areas, and therefore both Metric and Imperial are shown when practical. Canada and the United States have adopted similar but not entirely identical electrical codes. This book has attempted to build a link between the two countries and lists both CEC and NEC requirements. With the adoption of the North American Free Trade Agreement it is convenient and advantageous to be aware of each country's code requirements.

About the Author

Herb Putz has taught electrical technology for many years at a polytechnical institute. He is a master electrician, Registered Engineering Technologist and a teacher. His experience in electrical technology spans over 20 years, and includes installing and maintaining electrical equipment, designing electrical systems and teaching electrical technology.

ACKNOWLEDGEMENTS

The author and publisher express their sincere appreciation to those listed below for their assistance in the development of this publication:

Illustrations:
- Ian Holmes
- Ted Leach

Jasper Printing Group

Proofreading: A special thank you is extended to the following for their many hours spent proofreading this book:
- Werner Lemmer (P.Eng.)
- Robert Davis (CET)
- Marie Kennedy (Court Reporter, CSR (A))
- Brenda Putz (B. Admin.)

TABLE OF CONTENTS

TABLE OF CONTENTS

SECTION ONE
ONE
FUNDAMENTALS

Electrical Historical Notes

Historians credit the ancient Greeks with being the first to recognize and label electricity. The word electric is derived from a Greek word meaning amber. The Greeks discovered that if amber was rubbed with a cloth mysterious forces of attraction and repulsion occurred. It would be many centuries before there was any understanding of these forces.

The word *electricity* was first used by Sir Thomas Browne (1605-82) a noted English physician. Stephen Gray (1696-1736) discovered that some substances conduct electricity and others did not. Charles du Fay discovered two kinds of electricity that he called vitreous and resinous which are known today as positive and negative charges. Benjamin Franklin (1706-1790) performed his famous kite experiment to prove the "liquidity" of electricity. Charles Augustin de Coulomb (1736-1806) proposed laws that govern the behaviour of charged particles.

Luigi Galvani (1737-1798) worked and experimented with electricity. Alessandro Volta (1745-1827) discovered the chemical action that occurs between moisture and two different metals and constructed the first battery.

Hans Christian Oersted (1777-1836) discovered the electromagnetic effects of current flowing through a conductor. Andre Ampere (1775-1836) measured the magnetic effect of an electrical current and developed the fundamental laws that are basic to the study of electricity.

Georg Simon Ohm (1787-1854) produced the mathematical relationships among the three very important electrical quantities of resistance, voltage and current.

Michael Faraday (1791-1867) and Joseph Henry (1797-1878) independently discovered the principle of electromagnetic induction. James Clerk Maxwell (1831-1879) produced the first set of mathematical laws governing electricity and magnetism.

Electrical Historical Notes (cont'd)

Heinrich Rudolph Hertz (1857-1894) discovered electromagnetic waves.

Following these early scientists came others such as: Thomas Edison, Robert Millikan, John Fleming, Heinrich Lenz, Charles Steinmetz, Nikola Tesla, George Westinghouse, Werner Siemens, Gustav Kirchoff and many others.

Electricity Basics - Charges

Over the centuries scientists discovered that electricity is predictable and measurable. Being familiar with known rules, laws and formulas results in an understanding about how and why electrical circuits work. All matter is composed of atoms, and all atoms are composed of electrons, protons and neutrons. Illustration #1 indicates the charge that each of the components of an atom contain.

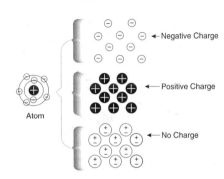

Illustration #1 - Electrical Charges of an Atom

Most matter exists in an uncharged state, meaning that there are equal numbers of positive and negative charges. However, this neutral state can be changed in a number of different ways. Illustration #2 shows six different means of altering the neutral state of matter.

Electricity Basics - Charges

+ Charges and Electrons
are Present in Equal
Quantities in the Rod and Fur

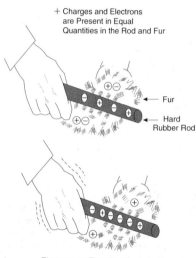

Fur

Hard
Rubber Rod

Electrons are Transferred
From the Fur to the Rod

Illustration #2A - Charged Particles From Friction

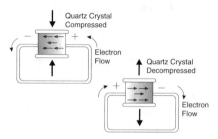

Quartz Crystal
Compressed

Electron
Flow

Quartz Crystal
Decompressed

Electron
Flow

**Illustration #2B - Piezoelectricity Charge Resulting from
Pressure on Quartz Crystals**

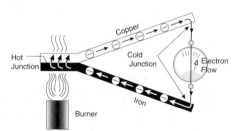

Copper

Hot
Junction

Cold
Junction

Electron
Flow

Iron

Burner

**Illustration #2C - Thermoelectricity - Heat at the Junction
of Two Dissimilar Metals**

Electricity Basics - Charges

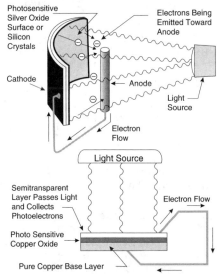

Photosensitive Silver Oxide Surface or Silicon Crystals

Electrons Being Emitted Toward Anode

Cathode

Anode

Light Source

Electron Flow

Light Source

Semitransparent Layer Passes Light and Collects Photoelectrons

Electron Flow

Photo Sensitive Copper Oxide

Pure Copper Base Layer

Illustration #2D - Photoelectricity - Unlike Charges From Light Impressed on Photosensitive Material

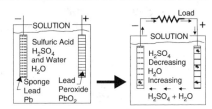

SOLUTION − +

Sulfuric Acid H_2SO_4 and Water H_2O

Sponge Lead Pb

Lead Peroxide PbO_2

Load

SOLUTION

H_2SO_4 Decreasing H_2O Increasing

$H_2SO_4 + H_2O$

Illustration #2E - Battery - Charged Particles from Two Dissimilar Materials

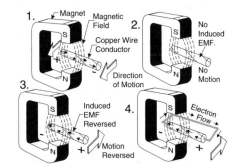

1. Magnet
Magnetic Field
Copper Wire Conductor
Direction of Motion

2. No Induced EMF.
No Motion

3. Induced EMF Reversed
Motion Reversed

4. Electron Flow

Illustration #2F - Generator - A Charge Produced by Moving a Conductor Through a Magnetic Field

Electricity Basics - Materials

There are three types of electrical materials:

1. Conductors: Materials that allow current to flow easily. Conductors have a large number of free electrons and from one to three valence electrons in their atomic structure. Most metals are conductors; silver, copper, gold and aluminum are ranked as the very best (see page 52, table #5).

2. Insulators: Materials that do not readily allow current to flow. Insulators are intended to prevent the flow of current. They have five to eight valence electrons in their atomic structure. Rubber, glass and mica are good insulators.

3. Semiconductors: Materials that may carry current, but not readily. They have four valence electrons in their atomic structure. Silicon and germanium are common semiconducting materials.

Note: See Section Two for a detailed examination of conductors and insulators.

Electrical Quantities

Electrical Current: An electron is approximately 2000 times lighter than a proton; therefore, the electron will travel or move rather than the proton. Electric current is a drift or movement of electrons, or negative charges, through a conducting medium. To quantify current a unit called "Coulomb" was adopted. One Coulomb is equal to 6.242×10^{18} electrons. One ampere is equal to one Coulomb per second (see illustration #3). Current is represented by the capital letter I, and its unit is the ampere or amp with the symbol A.

Electromotive Force (EMF): Because electrons are evenly distributed and do not exist naturally in a charged state, an electron moving force (EMF) is required to detach them from their atoms. This electron moving force or pressure is called a potential or voltage.

Electromotive Force (cont'd):

The unit adopted to measure this electrical pressure is the volt (V). The higher the voltage the higher the pressure to cause electrons to move. Illustration #2 shows six methods by which an EMF could be created.

Opposition: As electrons flow or move through a material they encounter opposition. The type of opposition encountered will depend on certain physical and electrical properties. Opposition offered by the molecular structure of a material is known as resistance, and it is measured in a unit called ohms (Ω).

Basic Electric Circuit: Illustration #4 compares an electrical circuit to that of a water piping system. Parallels can be drawn between the two for purposes of understanding and visualizing how an electrical circuit works. The electrical circuit must contain a source of pressure (EMF), a conducting medium (conductor) and a load.

Note: There are two conventions to show current flow. Conventional current flow is shown positive to negative, and electron current flow is shown negative to positive.

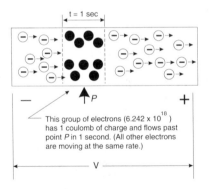

This group of electrons (6.242×10^{18}) has 1 coulomb of charge and flows past point P in 1 second. (All other electrons are moving at the same rate.)

Illustration #3 - The Unit "Ampere"

Basic Electric Circuit (cont'd):

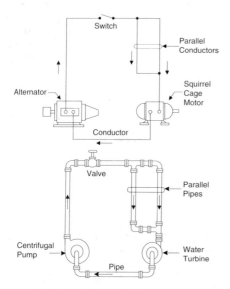

Illustration #4 - Basic Electric Circuit Analogy

Resistance

Electrons flowing through a material will encounter atoms and occasionally collide with them. The more atoms that the electrons encounter the more opposition there is to the flow of electrons. Each material type differs in the number of collisions that occur. The property of a material that restricts the flow of electrons is called resistance, and thus resistance can be defined as the opposition to current. All electric circuits have resistance, and when current flows through a resistive device heat is produced. The amount of resistance is dependent upon the following:

a. type of material (resistivity) (ρ)
b. cross-sectional area of material (A)
c. length of material (L)
d. temperature (T)

Based on an ambient temperature of 68°F (20°C), the resistance of a material can be calculated using the following formula:

$$R = \rho \times \frac{L}{A}$$

Resistance(cont'd)

Table #1 lists the resistivity (ρ) for various materials.

The Resistivity of Various Materials	
Material	Ω **per cmil-foot (ρ)**
Silver	9.9
Copper, annealed	10.37
Gold	14.7
Aluminum	17.0
Tungsten	33.0
Nickel	47.0
Iron, commercial	74.0
Constantan	295.0
Nichrome	600.0
Calorite	720.0
Carbon	21,000.0

Table #1 – Resistivity (ρ)

Resistance Calculation - Example #1

Calculate the resistance of a copper conductor with a cross-sectional area (A) of 4000 circular mils and a length (L) of 400 feet.

$R = \rho \times L/A$

$R = 10.37 \times {}^{400}/_{4000}$ $R = 1.037 \ \Omega$

Resistance Calculation - Example #2

Calculate the resistance of a aluminum conductor with a diameter of ⅛ of an inch and a length of 800 feet.

⅛ inch = 125 mils (${}^{125}/_{1000}$)

circular mils = 125^2 = 15625

$R = 17 \times {}^{800}/_{15625}$

$R = 0.87 \ \Omega$

Note: The American Wire Gage (AWG) is based on Imperial units, and therefore the calculations shown here are in Imperial units. The area of a round conductor for purposes of the resistance formula is based on the mil which is equal to $^1/_{1000}$ of an inch. To calculate the area of a round conductor simply express the diameter in mils and then square the value. Refer to Section Two, page 78 for more details.

Resistance (cont'd)

Temperature has a significant effect on conductor resistance. Table #2 lists the temperature coefficients (α) of various materials based on an ambient temperature of 20°C.

Temperature Coefficient of Resistance for Various Conductors at 20°C	
Material	**Temperature Coefficient (α)**
Silver	0.0038
Copper, annealed	0.00393
Gold, pure	0.0034
Aluminum	0.00391
Tungsten	0.0045
Nickel	0.006
Iron, commercial	0.0055
Constantan	0.000008
Nichrome	0.00044

Table #2 - Temperature Coefficient of Resistance

Resistance can be determined for a conductor at any temperature by the use of the following formula:

$$R_2 = R_1[1 + \alpha (T_2\text{-}20)]$$

R_1 = Value of Resistance at 20°C
R_2 = Resistance at T_2
α = Temperature Coefficient from table #2
T_2 = New Ambient Temperature (°C) of the conductor

Resistance Calculation - Example #3

Calculate the new resistance value of a 10 Ω silver conductor if the ambient temperature is raised to 60°C.

$$R_2 = R_1[1 + \alpha (T_2\text{-}20)]$$
$$R_2 = 10 [1 + 0.0038(60\text{-}20)]$$
$$R_2 = 11.52 \ \Omega$$

Resistors

Components that are specifically designed and manufactured to have a certain amount of resistance are called resistors. There are two types of resistors: fixed and variable. Illustrations #5A and #5B provide an example of a fixed and a variable resistor.

Fixed resistors are available in a variety of sizes and are rated in both ohms and watts. Fixed resistors with value tolerances of 5 - 20% are color coded with four bands to indicate their values of resistance and tolerance. Illustration #6 shows the banding, and table#3A lists the values associated with each color.

A fifth band may appear on some resistors indicating the reliability of the resistor expressed as a percentage of failures during 1000 hours of operation. Table #3B lists the reliability color code.

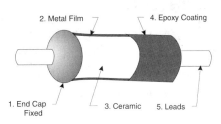

2. Metal Film 4. Epoxy Coating

1. End Cap Fixed 3. Ceramic 5. Leads

Illustration #5A - Fixed Resistor

R

Illustration #5B - Variable Resistor

Resistors (cont'd)

COLOR BAND VALUES		
	Digit	**Color**
Resistance value first three bands,	0	Black
	1	Brown
	2	Red
1st band	3	Orange
- 1st digit	4	Yellow
2nd band	5	Green
- 2nd digit	6	Blue
	7	Violet
3rd band	8	Gray
- number of zeros	9	White
(For multiplying band only)	0.1	Gold
	0.01	Silver
Tolerance, fourth band	5%	Gold
	10%	Silver
	20%	No band

Table #3A - Color Band Values

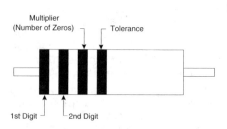

Illustration #6 - Resistor Color Bands

Fifth Band Reliability Color Code	
Color	**Failures (%) during 1000 hours of operation**
Brown	1.0%
Red	0.1%
Orange	0.01%
Yellow	0.001%

Table #3B - Reliability Color Code

Resistors (cont'd)

Calculation of fixed resistors

Determine the resistance value of the resistor shown in illustration #7.

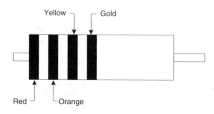

Illustration #7 - Resistor Example

1. first band is red = 2
2. second band is orange = 3
3. third band is yellow = 4 zeros
4. fourth band is gold = 5% tolerance

Therefore this resistor has a resistance value of 230,000 Ω accurate to within $\pm$ 5% of the calculated value.

Potentiometers (Variacs), Rheostats

Variable resistors are used to manually or automatically change a resistance value. Potentiometers or variacs are used to vary the source voltage as shown in illustration #8A. Rheostats are used to vary the resistance values of an electrical circuit, as shown in illustration #8B.

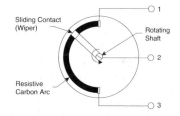

Illustration #8A - Basic Potentiometer/Variac Construction
Three Lead Device

Resistors (cont'd)

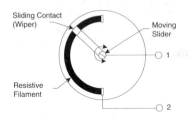

Illustration #8B - Basic Rheostat Two Lead Device

Thermistors, Varistors, Photoconductive Cells Thermistors are a special type of resistor made from semi-conducting metallic oxide conductors. These devices are primarily used in measurement applications in the field of instrumentation. Varistors are zinc oxide or silicon carbide crystals that act as resistors until a sudden change in voltage takes place, at which time they become excellent conductors.

Photoconductive cells (photoresistors) are variable resistors sensitive to light.

Thermistor: resistance decreases as temperature increases.

Varistor: resistance decreases as applied voltage increases.

Photoresistor: resistance decreases as incident light increases.

Ohm's Law

The amount of current flowing in an electrical circuit is dependent upon the value of electrical pressure (E) and the amount of opposition to the flow of current (R). A mathematical formula representing this relationship reveals that:

$$I = E/R$$

Note: This is known as Ohm's Law. It is a fundamental formula for the determination of the behaviour of an electrical circuit.

Ohm's Law (cont'd)

Ohm's Law Calculation Examples:

Calculate the amount of current flowing in a circuit having $100\,\Omega$ of opposition and 120 V of pressure.

$I = E/R$

$I = 120/100$

$I = 1.2$ A

Calculate the resistance of a circuit that allows only 0.5 A to flow when a voltage of 120 V is applied.

$R = E/I$

$R = 120/0.5$

$R = 240\ \Omega$

Power and Energy

Power (P) is defined as the rate at which work is being done or energy is being used and is expressed in an electrical unit called watts. Energy is power consumed over a period of time, and its unit is kilowatt hours (kWh).

Whenever current flows through an electrical circuit, power is consumed. The amount of power consumed is dependent upon the current and the resistance and can be mathematically formulated:

Power (P) = Voltage across resistor (V) x Current through resistor (I)

$$P = V\ x\ I$$

As V = I x R, we may substitute for V in the power formula, and the result will be:

$$P = I^2\ x\ R$$

If, in the power formula, V/R is substituted for I the resultant power formula will be:

$$P = V^2/R$$

Illustration #9 summarizes the Ohm's Law mathematical relationships. The inner circle represents the circuit value being looked for and the outer circle provides the formulas that are applicable.

Power and Energy (cont'd)

Electrical energy consumption may be determined from the formula:

Energy (E) = Power (P) x Time (T)

and is usually expressed in kilowatt hours (kWh).

Note: The quantity symbol for Energy is actually W; however, this is often confused with the unit symbol for Power. Therefore, in this book E will be used to symbolize Energy.

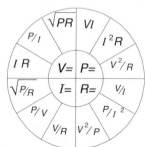

Illustration #9 - Ohm's Law Relationships

To determine the cost of operating an electrical appliance simply multiply the cost per kWh by the number of kWh. For example, the cost to operate a 15 kW dryer for 4 hours if the energy cost is $ .07 per kWh is: 15 x 4 x .07 = $4.20.

Basic Electrical Formulas:

$I = V/R$ - Ohm's law

$V = I R$ - Ohm's law

$R = V/I$ - Ohm's law

$P = I^2 R$ - Power equals current squared times resistance

$P = V I$ - Power equals voltage times current

$P = V^2/R$ - Power equals voltage squared divided by resistance

$P = E/t$ - Power equals energy divided by time

$E = P t$ - Energy equals power multiplied by time

Series Circuits

In a series circuit there is only one path for the current to flow; therefore there can only be one value of current. In a resistive series circuit, resistors are connected end to end or in a string configuration as shown in illustration #10.

The total opposition of a series circuit is obtained by adding all the resistors together. Illustration #11 shows three resistors connected in series to a voltage source. These three resistors are equivalent to one resistor with a value equal to the sum of the three resistors shown.

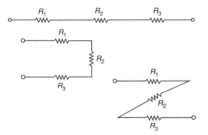

Illustration #10 - Resistive Series Circuits

All of the circuits shown in illustration #10 are connected in series even though their presentation differs somewhat.

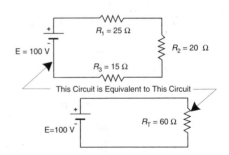

Illustration #11 - Series Equivalent Circuit

Series Circuits (cont'd)

Note: The current that flows through all three resistors shown in illustration #11 has one value. The voltage drop across each of the resistors of illustration #11 is equal to the current through the resistor times the ohmic value of the resistor. The sum of the individual voltage drops will equal the voltage applied by the source. Illustration #12 shows how the above mentioned relationships work in a series circuit.

The following summarizes the formulas for a series circuit:

Series Circuit Formulas

$$R_t = R_1 + R_2 + R_3 + ...$$
$$I_t = I_1 = I_2 = I_3 = ...$$
$$V_t = V_1 + V_2 + V_3 + ...$$
$$P_t = P_1 + P_2 + P_3 + ...$$

Series Circuit Calculation Example

Calculate the following values for the circuit shown in illustration #13.

a. total resistance

b. total power

c. current through R_2

d. voltage drop across R_3

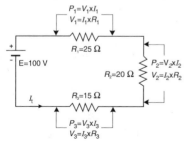

$$P_1 = V_1 \times I_1$$
$$V_1 = I_1 \times R_1$$
$$R_1 = 25\ \Omega$$
$$E = 100\ V$$
$$P_2 = V_2 \times I_2$$
$$V_2 = I_2 \times R_2$$
$$R_2 = 20\ \Omega$$
$$R_3 = 15\ \Omega$$
$$I_1$$
$$P_3 = V_3 \times I_3$$
$$V_3 = I_3 \times R_3$$

Illustration #12 - Series Circuit Relationships

Series Circuits (cont'd)

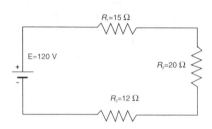

Illustration #13 - Calculation Example

Example Solution:

a. $R_t = R_1 + R_2 + R_3$
   $R_t = 15 + 20 + 12$ $R_t = 47\ \Omega$

b. $P_t = P_1 + P_2 + P_3$ or
   $P_t = V_t \times I_t$ or
   $P_t = V_t^2 / R_t$
   $P_t = 120^2/47$ $P_t = 306.4$ W

c. $I_t = I_1 = I_2 = I_3$ or
   $I_t = V_t / R_t$
   $I_t = 120/47$ $I_t = I_2 = 2.55$ A

d. $V_3 = I_3 \times R_3$
   $V_3 = 2.55 \times 12$ $V_3 = 30.6$ V

Voltage Divider Formula

Although Ohm's Law can be used to determine the voltage drop across any resistor, it is necessary to first know the current through the resistor. By using the voltage divider formula, a value may be obtained without having to first calculate the current.

The voltage drop across any resistor or combination of resistors in a series circuit is equal to the ratio of that resistance value to the total resistance multiplied by the source voltage.

Illustration #14 shows the formula and how it applies to a series circuit.

Voltage Divider Formula (cont'd)

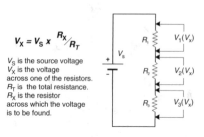

$$V_X = V_S \times \frac{R_X}{R_T}$$

V_S is the source voltage
V_X is the voltage
across one of the resistors.
R_T is the total resistance.
R_X is the resistor
across which the voltage
is to be found.

Illustration #14 - Voltage Divider Formula

Parallel Circuits

When two or more electrical components are connected across the same voltage source they are considered to be in parallel. A parallel circuit offers more than one path for current to flow, and each of these paths is called a branch.

Most electrical wiring for residential, commercial and industrial plants are in a parallel configuration. Luminaires (fixtures), motors, heaters, appliances, receptacles, etc. are all connected in parallel. Illustration #15 shows various circuits connected in parallel.

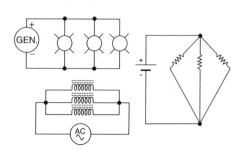

Illustration #15 - Parallel Circuits

Parallel Circuits (cont'd)

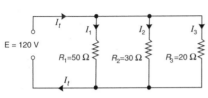

Illustration #18 - Parallel Circuit

Parallel Circuit Calculations:

Based on the circuit shown in illustration #18, calculate for the following:

a. I_1 d. I_t
b. I_2 e. R_t
c. I_3 f. P_t

Parallel Circuit Solutions:

a. $I_1 = V_1/R_1$ $I_1 = 120/50$ $I_1 = 2.4$ A

b. $I_2 = V_2/R_2$ $I_2 = 120/30$ $I_2 = 4.0$ A

c. $I_3 = V_3/R_3$ $I_3 = 120/20$ $I_3 = 6.0$ A

d. $I_t = I_1 + I_2 + I_3$ $I_t = 12.4$ A

e. $R_t = V_t/I_t$ or $1/R_t = 1/R_1 + 1/R_2 + 1/R_3$
$R_t = 120/12.4$
$R_t = 9.68 \ \Omega$
$1/R_t = 1/R_1 + 1/R_2 + 1/R_3$
$1/R_t = 1/50 + 1/30 + 1/20$ $1/R_t = 0.1033$
taking the reciprocal of both sides yields:
$R_t = 1/0.1033$ $R_t = 9.68 \ \Omega$

f. $P_t = V_t \times I_t$ or $P_t = I_t^2 \times R_t$
or $P_t = V_t^2/R_t$ or $P_t = P_1 + P_2 + P_3$
$P_t = V_t \times I_t$ $P_t = 120 \times 12.4$ $P_t = 1488$ W
$P_t = I_t^2 R_t$ $P_t = 12.4^2 \times 9.68$ $P_t = 1488$ W
$P_t = V_t^2/R_t$ $P_t = 120^2/9.68$ $P_t = 1488$ W
$P_t = P_1 + P_2 + P_3$
$P_t = (V_1 \times I_1) + (V_2 \times I_2) + (V_3 \times I_3)$
$P_t = (120 \times 2.4) + (120 \times 4) + (120 \times 6)$
$P_t = 1488$ W

Parallel Circuits (cont'd)

The Current Divider Principle is useful in quickly determining the value of current flowing through any of the individual branches. Illustration #19 shows how the formula $Ix = It \times \frac{Rt}{Rx}$ can be used to calculate the current in each branch.

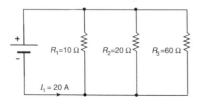

Illustration #19 - Current Divider

$1/R_t = \frac{1}{10} + \frac{1}{20} + \frac{1}{60}$ $R_t = 6\ \Omega$

$I_1 = 20 \times R_t / R_1$ $I_1 = 20 \times 6/10$ $I_1 = 12$ A

$I_2 = 20 \times R_t / R_2$ $I_2 = 20 \times 6/20$ $I_2 = 6$ A

$I_3 = 20 \times R_t / R_3$ $I_3 = 20 \times 6/60$ $I_3 = 2$ A

In summary, the following statements may be made about a parallel circuit:
1. Equal voltages exist across each branch.
2. The total resistance is always less than the least individual branch resistance.
3. The total current in the circuit is equal to the sum of the individual branch currents. The junction point of these currents is called a node, and Kirchoff's current law states that the sum of the currents entering a node is equal to the sum of the currents leaving the node.
4. The highest current appears in the branch with the lowest resistance, and the lowest current appears in the branch with the highest resistance.
5. The total power in the circuit is equal to the sum of the individual powers.
6. When one of the resistors in a parallel circuit is removed, the total resistance of the circuit will increase.

Series-Parallel Circuits

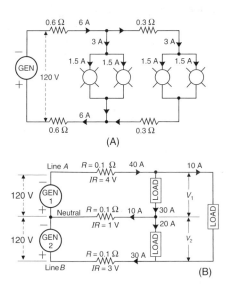

(A)

(B)

Many common electrical circuits are connected in a series-parallel configuration, as shown in illustration #20.

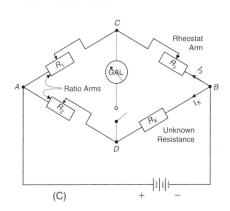

(C)

Illustration #20 - Series-Parallel Circuits

Series-Parallel Circuits (cont'd)

Sectioning of the total circuit should be the first step in solving circuit values. Dividing the complex circuit into a series of smaller circuits simplifies the task. Examine the entire circuit, and determine which parts of the circuit are connected in parallel and which parts are connected in series. Then begin to reduce these parts into simpler forms by representing them as equivalent values.

After the circuit has been reduced to its simplest form, all other values may be obtained. The following calculations are based upon the circuit shown in illustration #21.

Series-Parallel Calculations

Calculate the total equivalent resistance and the total current for the circuit shown in illustration #21.

Series-Parallel Solution

Reduce the circuit from the complex network shown in illustration #21 into the simpler circuits shown in illustration #22. Once the circuit is reduced into its simplest form, calculate for R_t and I_t.

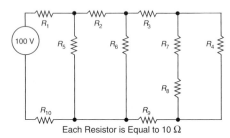

Each Resistor is Equal to 10 Ω

Illustration #21 - Series-Parallel Example

Series-Parallel Circuits (cont'd)

Step #1: Reduce the circuit beginning from the extreme right side and working to the left. R_7 & R_8 are in series, therefore $R_{(7)(8)} = 10 + 10 = 20\ \Omega$

Step #2: New equivalent values.

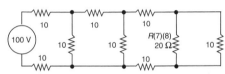

Step #3: $R_{(7)(8)}$ is in parallel with R_4.
$$R_{(4)(7)(8)} = \frac{(20 \times 10)}{(20 + 10)} = 6.67\ \Omega$$

Step #4: New equivalent values.

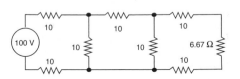

Step #5: $R_{(3)}$, $R_{(9)}$, $R_{(4)(7)(8)}$ are all in series. $R_{(3)(4)(7)(8)(9)} = 10 + 6.67 + 10 = 26.67\ \Omega$

Step #6: New equivalent values.

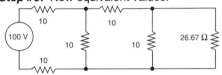

Step #7: R_6 is in parallel with $26.67\ \Omega$.
$$R_{(3)(4)(6)(7)(8)(9)} = \frac{(10 \times 26.67)}{(10 + 26.67)} = 7.3\ \Omega$$

Step #8: New equivalent values.

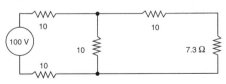

Illustration #22 - Reducing a Complex Circuit

Series-Parallel Circuits (cont'd)

Step #9: R_2 is in series with 7.3 Ω.
$R_{(2)(3)(4)(6)(7)(8)(9)} = 10 + 7.3 = 17.3Ω$

Step #10: New equivalent values.

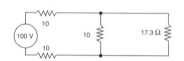

Step #11: R_5 is in parallel with 17.3 Ω.
$R_{(2)(3)(4)(5)(6)(7)(8)(9)} = \dfrac{(10 \times 17.3)}{(10 + 17.3)} = 6.34\ Ω$

Step #12: New equivalent values.

Step #13: $R_{(1)}$ and $R_{(10)}$ are in series with 6.34Ω.

$R_{equivalent} = 10 + 6.34 + 10 = 26.34\ Ω$

Step #14: $I_t = V_t / R_{equivalent}$
$I_t = {}^{100}/_{26.34}$ $I_t = 3.79\ A$

Three-Wire Edison

The circuit arrangement shown in illustration #20B is commonly used in residential wiring and offers the option of connecting a load to one of two voltages. Lighting and general convenience receptacles are usually connected to 120 V; whereas, the stove, electric hot water heater, and air conditioner are connected to 240 V.

The wire that is common to the two line wires is known as the neutral. The neutral wire carries the unbalanced current and is usually the same AWG size as either one of the line conductors. There are several advantages to having an electrical service at two voltage levels: lower line losses, lower line drops, smaller conductors and two levels of voltage should a load require it.

Three-Wire Edison (cont'd)

As indicated in illustration #20B, the neutral wire carries the unbalanced current. This unbalanced current is only present if the two line currents are different. A potentially serious condition may occur if the neutral becomes open circuited. This condition, referred to as an open neutral, may result in severe voltage unbalances between loads and may damage or destroy the electrical circuit. The neutral conductor should not be switched, fused or protected by a circuit breaker.

Note: In residential services the neutral conductor is always grounded and is identified by a white color.

It should be noted that not all white colored conductors are considered to be neutral. A conductor may only be considered to be neutral if it carries the **unbalanced current**. If the white conductor is not a true neutral, it is then called the identified circuit conductor or the grounded circuit conductor.

Illustration #23 provides an example of circuits where the white conductor is the neutral and circuits where the white conductor is the grounded circuit conductor.

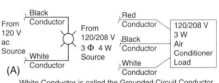

White Conductor is called the Grounded Circuit Conductor

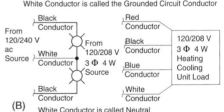

(B)   White Conductor is called Neutral

Illustration #23 - White Colored Conductors

Wheatstone Bridge

Illustration #20C, a Wheatstone Bridge circuit, may be redrawn as shown in illustration #24. This bridge circuit is commonly used in measurement devices such as an ohmmeter or for temperature measurement involving a temperature sensitive element such as a thermistor. The advantage of this bridge circuit is in its simplicity for comparing an unknown circuit value to those already known.

As illustration #24 shows, the output terminals are connected to a measuring device called a galvanometer. When an unknown resistance value (R_x) is connected across terminals B and D, adjustments to the rheostat will eventually result in a zero deflection to the center scale galvanometer. The value of the unknown resistor can then be determined based on the ratios and values of the other resistors.

The following formula may be used to determine the value of the unknown resistor (Rx):

$$R_x = R_2 \ x \ R_3 \ / \ R_1$$

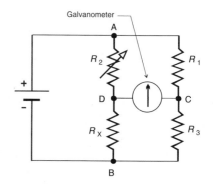

Illustration #24 - Wheatstone Bridge Circuit

Alternating Current

The earliest sources of electrical energy came from DC sources such as batteries. However, in a direct current system there are some inherent limitations and disadvantages which have spurred on the growth of AC systems. Over 90% of modern day generated electrical energy is AC.

The following list summarizes some of the advantages of AC over DC:

1. The generated voltage and the distribution voltage do not have to be the same.

2. The transmission voltage can be at very high values, therefore both line losses and line drop are kept to a minimum.

3. AC voltages are easily transformed from one voltage level to another.

4. AC voltages may be varied electronically using SCRs (Silicon Controlled Rectifiers) or Triacs.

5. AC motors and alternators do not require commutators and are therefore not limited in size and speed.

6. AC motors are smaller, simpler and more efficient than DC motors.

An alternating voltage is produced by rotating a magnetic field past a set of stationary conductors as shown in illustration #25.

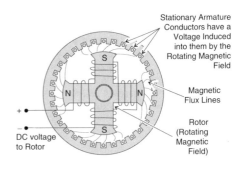

Stationary Armature Conductors have a Voltage Induced into them by the Rotating Magnetic Field

Magnetic Flux Lines

Rotor (Rotating Magnetic Field)

DC voltage to Rotor

Illustration #25 - Generating an AC Voltage

Alternating Current (cont'd)

The waveform produced by an alternator changes in magnitude and polarity. This waveform is referred to as a sine wave because the magnitude of the waveform at any instant in time can be calculated by multiplying the peak value by the sine of the angle:

$$e = E_{max} \times \sin \theta.$$

The sine wave is also referred to as a sinusoidal wave and consists of two alternations which make up one cycle. The number of cycles that occur in one second is referred to as the frequency and is expressed in Hertz (Hz). In North America the frequency of a generated AC voltage is 60 Hz.

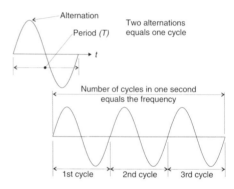

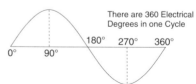

Illustration #26 - AC Cycles

Alternating Current - RMS Values

AC voltages are always expressed as RMS values. The term RMS represents Root-Mean-Square, and refers to a mathematical process followed to determine the effectiveness of AC as compared to DC. The RMS value of a sine wave is a measure of its AC heating effect (I^2R) as compared to that of a similar DC value.

The RMS value of a sine wave may be calculated by multiplying the peak AC value by 0.707, or conversely, the RMS value may be determined by dividing the AC peak value by 1.414.

Note: All future references to AC values will be in RMS.

Not all waveforms in AC power systems are sinusoidal; some waveforms are nonsinusoidal as shown in illustration #27B. A repetitive nonsinusoidal waveform is composed of a fundamental waveform and a series of multiples of the fundamental.

The sum of these waveforms results in a nonsinusoidal waveform. If a waveform varies from being a pure sine wave, it then contains harmonics. Harmonics, as shown in illustration #27B, are waveforms that are multiples of the fundamental in both frequency and amplitude.

A third harmonic has a frequency that is three times that of the fundamental and usually a magnitude that is one-third that of the fundamental. A fifth harmonic has a frequency that is five times that of the fundamental and a magnitude that is one-fifth that of the fundamental. The sum of these values results in a composite nonsinusoidal waveform.

Harmonics are a growing concern to power users and power suppliers because they result in unwanted side effects such as heating of feeder conductors, heating of rotors in motors and interference in signal transmissions.

Alternating Current - RMS Values (cont'd)

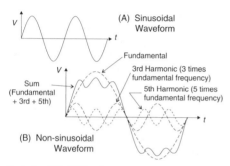

Illustration #27 - AC Waveforms

Inductance - Magnetic Field

An understanding of the electrical property called inductance requires a fundamental knowledge of magnetism. There is a close relationship between the flow of electric current and magnetism.

In 1819, Hans Christian Oersted observed that a magnetic field is formed whenever current flows through a conductor. There is no physical link between the conductor and the magnetic field, and the magnetism is due to the motion of electrons in the atoms. The following statements can be made regarding magnetic circuits and their properties:

1. A magnetic field is that region in which a magnetic material is acted upon by a magnetic force.

2. Magnetic lines of force have direction. The magnetic lines of force always leave the north pole and enter the south pole. See illustration #28A.

3. Magnetic lines of force form complete circuits or loops. See illustration #28.

4. Like poles repel and unlike poles attract. See illustration #28C.

5. Magnetic lines of force do not cross or intersect.

Inductance - Magnetic Field (cont'd)

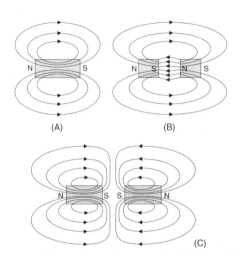

(A)   (B)

(C)

Illustration #28 - Magnetic Fields

Whenever electrical current flows, a magnetic field called an electromagnetic field is produced. Magnetic lines of force form concentric patterns around an electrical conductor as indicated in illustration #29A. The lines of force are strongest near the conductor. The direction of the magnetic lines of force is dependent upon the direction of the current flow and can easily be determined using the left hand rule for electron current flow or the right hand rule for conventional current flow. See illustration #29B.

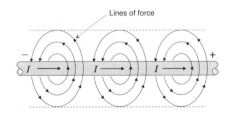

Illustration #29A - Magnetic Lines of Force (Electron Flow)

Inductance - Magnetic Field (cont'd)

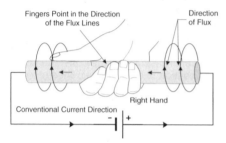

Illustration #29B - Right Hand Rule for a Conductor

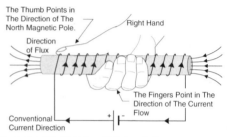

Illustration #30 - Right Hand Rule for a Coil

Concentrating the magnetic lines of force by winding a conductor around a magnetic material such as iron will cause the magnetic field strength to increase. An electromagnet, used in solenoids and relays, may be made by winding an electrical conductor made of copper around a bar of iron. The magnetic polarity may be determined by using the right hand rule for coils as shown in illustration #30.

Electromagnetic Induction and Induced Voltage

If a conductor is moved past a magnetic field, a voltage will be induced into that conductor. This action is called electromagnetic induction, and the resulting voltage is called an induced voltage. The amount of voltage that is induced into the conductor is dependent upon the following factors:

Electromagnetic Induction (cont'd)

1. The amount of magnetic flux or lines of force that are present.
2. The rate at which the field moves past the conductor or the conductor moves past the field.
3. The number of turns of wire in the winding of the conductor.

Inductance becomes the property of any electrical circuit that experiences a changing current and is dependent upon the number of turns of wire (N), the permeability (μ) (ease to magnetize) of the core material, the area (A) of the magnetic circuit and the length(l) of the magnetic circuit (the longer it is, the weaker it becomes). Inductance is expressed in Henries and symbolized with the capital letter (L).

$$L = N^2 \mu A / l$$

Self-Inductance: Self-inductance occurs whenever a changing current flows through a conductor and creates a magnetic field which, in turn, induces a voltage on to itself. This voltage is called a Counter EMF (CEMF) and opposes the voltage that created it.

Note: Lenz's Law summarizes this principle by stating that "the induced EMF in any circuit is always in a direction to oppose the effect that produced it".

In an AC circuit the current is constantly changing in magnitude and polarity. The amount of EMF that is induced in any circuit is dependent upon this rate of change of the current (magnetic field). This principle of self inductance is important in understanding how current is limited in AC inductive circuits.

Mutual Inductance: Mutual inductance occurs whenever a conductor that is adjacent to a current carrying conductor has a voltage induced into it.

Electromagnetic Induction (cont'd):

A magnetic field is created by the current carrying conductor, and it is this field that cuts across the adjacent conductor and induces a voltage into it. All previously mentioned factors for self-inductance still apply. Mutual inductance is what makes a transformer work. There must be a constantly changing current in order to have a constantly changing magnetic field. A changing magnetic field is equivalent to moving a conductor through a magnetic field and thereby inducing a voltage.

Series Inductors: Inductors that are connected in series will have their inductances added in the same manner that the resistances of resistors in series are added.

$$L_T = L_1 + L_2 + L_3 + ...$$

Parallel Inductors: Inductors that are connected in parallel will have their inductances added in the same manner that the resistances of resistors in parallel are added.

$$1/L_T = 1/L_1 + 1/L_2 + 1/L_3 ...$$

Current Lagging Voltage: Inductance is a characteristic of an electrical circuit that makes itself evident by opposing the starting, stopping or changing of current. This opposition to the change in current results in the current lagging the voltage by 90 electrical degrees in a purely inductive AC circuit as shown in illustration #31.

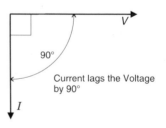

Current lags the Voltage by 90°

Illustration #31 - Current Lagging Voltage

Inductive Reactance

Inductive Reactance is a property of an AC circuit and is dependent upon the amount of inductance (L) in the circuit and the rate of change expressed as the frequency (f) of the electrical source. The greater the inductance (L), or the higher the frequency (f), the greater the inductive reactance of the AC circuit.

Inductive reactance opposes the flow of current in an AC circuit and is expressed in units of ohms. Inductive reactance (X_L) causes the current to lag the voltage by 90 degrees and is called an out of phase component.

The formula for inductive reactance is

$$X_L = 2\pi \ f \ L .$$

Current and voltage do not lag from each other in a purely resistive circuit but do lag by 90 degrees in a purely inductive circuit.

Illustration #32 shows electrical phasors (lines that indicate magnitude and direction) that represent the relationships of voltage and current for a purely resistive circuit, a purely inductive circuit, and a circuit that has both inductance and resistance.

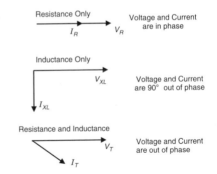

Resistance Only — Voltage and Current are in phase

Inductance Only — Voltage and Current are 90° out of phase

Resistance and Inductance — Voltage and Current are out of phase

Illustration #32 - Phase Relationships

Impedance

Impedance is defined as the total AC opposition to the flow of current. If both inductance and resistance are present in an AC circuit, then the phasor sum of the values must be obtained. Illustration #33 outlines the steps taken to obtain individual values in a circuit.

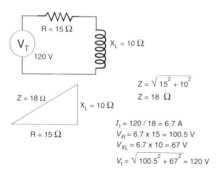

$$Z = \sqrt{15^2 + 10^2}$$
$$Z = 18 \ \Omega$$

$I_t = 120 / 18 = 6.7 \ A$
$V_R = 6.7 \times 15 = 100.5 \ V$
$V_{XL} = 6.7 \times 10 = 67 \ V$
$V_t = \sqrt{100.5^2 + 67^2} = 120 \ V$

Illustration #33 - Complex AC Circuit

Capacitance

A capacitor is a simple device that consists of two conductors separated by an insulating material as shown in illustration #34.

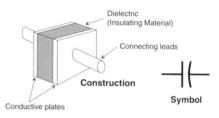

Illustration #34 - Simple Capacitor

Capacitance is a property of a circuit which opposes a change in voltage. Therefore, in an AC capacitive circuit the voltage lags the current by 90 electrical degrees. In a capacitive circuit, energy is stored in an electrostatic field by altering the orientation of electrons in their orbital path.

Capacitance (cont'd)

Illustration #35 shows both a discharged and charged capacitor and the orientation of their electrons.

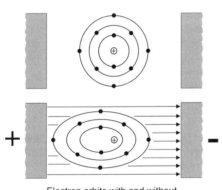

Electron orbits with and without
the presence of an electric field

Illustration #35 - Discharged & Charged Capacitor

Energy is stored by the distorted electron orbit and is released when the capacitor is discharged. The ability to store an electrical charge depends upon the capacitor's size. The farad (F) is the basic unit of capacitance, but in practicality the $\mu F(1 \times 10^{-6})$ is used (one farad is too large a unit). The capacity of a capacitor is dependent upon the following factors:

1. Area of the capacitor's plates (A). The larger the plate area the larger the capacitance value.

2. The ability of the dielectric material to concentrate lines of force between the plates (κ). See table #4 for a listing of both the dielectric constant and the dielectric strength.

3. The distance between the conducting plates (d). The smaller this distance the larger the capacitance.

In summary, the formula for capacitance is:

$$C \approx A \times \kappa/d$$

Capacitance (cont'd)

Dielectric Strength	
Average Dielectric Strength (kV / mm)	
Air	3
Barium-strontium titanate (ceramic)	3
Porcelain (ceramic)	8
Transformer Oil (organic liquid)	16
Bakelite (plastic)	16
Paper	20
Rubber	28
Teflon (plastic)	60
Glass	120
Mica	200
Note: These values refer to the materials' abilities to withstand electrical pressure (volts)	

Table #4A - Dielectric Strength

Dielectric Constants (κ)	
Air	1.0006
Barium-strontium titanate (ceramic)	7500
Porcelain (ceramic)	6
Transformer Oil (organic liquid)	4
Bakelite (plastic)	7
Paper	2.5
Rubber	3
Teflon (plastic)	2
Glass	6
Mica	5
Note: These values are used in the capacitance formula	

Table #4B - Dielectric Constants

Capacitors are classified according to the type of dielectric material used. The most common types of dielectric materials are mica, paper, plastic, ceramic and an electrolytic material composed of aluminum oxide and tantalum oxide.

When capacitors are connected in series, as shown in illustration #36A, their total capacitance decreases because the effective distance between their plates has increased. When capacitors are connected in parallel, as shown in illustration #36B, their total capacitance increases because the effective area of their plates has increased.

Capacitance (cont'd)

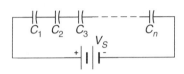

$$\frac{1}{C_T} = \frac{1}{C_1} + \frac{1}{C_2} + \frac{1}{C_3} + \ldots + \frac{1}{C_n}$$

$$C_T = \frac{1}{(1/C_1) + (1/C_2) + (1/C_3) + \ldots + (1/C_n)}$$

Illustration #36A - Capacitors in Series

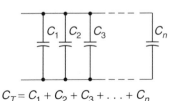

$$C_T = C_1 + C_2 + C_3 + \ldots + C_n$$

Illustration #36B - Capacitors in Parallel

A capacitor behaves in very predictable ways when connected into an electrical circuit. Illustration #37A shows that a capacitor when first being charged, or whenever experiencing a change in the charging pressure (voltage), acts like a short circuit. In illustration #37B, the capacitor is fully charged and behaves like an open circuit.

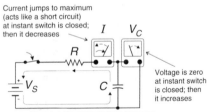

Current jumps to maximum (acts like a short circuit) at instant switch is closed; then it decreases

Voltage is zero at instant switch is closed; then it increases

Charging: Capacitor voltage increases as the current and resistor voltage decrease

Illustration #37A - Capacitor Being Charged

Capacitance (cont'd)

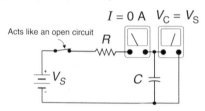

$$I = 0 \text{ A} \quad V_C = V_S$$

Acts like an open circuit

R

V_S

C

Fully charged: Capacitor voltage equals source voltage. The current is zero

Illustration #37B - Fully Charged Capacitor

Capacitive Reactance

This is an AC value and is dependent upon both the capacitance of the circuit and the frequency. Capacitive reactance is an opposition to the flow of current and is therefore expressed in ohms. The formula for capacitive reactance (X_C) is $X_C = 1 / (2 \pi f C)$

C is expressed in farads, and since a farad is too large a unit the μF shall be used in the formula.

This alters the formula into the format:

$$X_C = 10^6/(2 \pi f C),$$
$$f = frequency, \quad C = \mu F$$

As shown in illustration #38A, in a capacitive circuit the voltage lags the current by 90 electrical degrees. The circuit properties of resistance, inductive reactance and capacitive reactance are all out of phase (electrical time) with each other and can be represented by the phasors shown in illustration #38B.

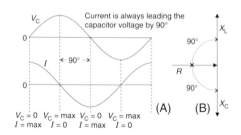

Current is always leading the capacitor voltage by 90°

V_C

I ← 90° →

$V_C = 0$ $V_C = $ max $V_C = 0$ $V_C = $ max
$I = $ max $I = 0$ $I = $ max $I = 0$

(A)   (B)

X_L
90°
R
90°
X_C

Illustration #38 - Capacitive Circuit

RLC Series Circuits

AC circuits which contain resistance, inductance and capacitance connected in series are known as RLC series circuits. In illustration #39, a series RLC circuit is shown.

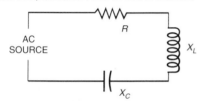

Illustration #39 - RLC Series Circuit

The following formulas may be used to obtain circuit values:

S (Apparent Power) = $V_t \times I_t$ or $I_t^2 \times Z$ or V_t^2/Z the unit is VA (volt amps).

P (True Power) = $V_R \times I_R$ or $I_R^2 \times R$ or V_R^2/R the unit is W (Watts).

Q (Reactive Power) = $V_X \times I_X$ or $I_X^2 \times X$ or V_x^2/X the unit is VARS (volt amps reactance).

Power Factor = P/S or R/Z or V_R/V_T there is no unit.

$Z = \sqrt{R^2 + (X_L - X_C)^2}$ the unit is ohms.

RLC Series Example Problem:

Calculate the following circuit values based upon the diagram in Illustration #40:

1. Z
2. I_t
3. V_R
4. V_{XL}
5. V_{XC}
6. True Power(P)
7. Apparent Power(S)
8. Reactive Power(Q)
9. Power Factor(PF)

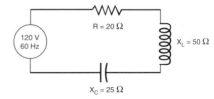

Illustration #40 - RLC Series Problem

RLC Series Circuits (cont'd)
RLC Series Example Solution:

1. $Z_T = \sqrt{R^2 + (X_L - X_C)^2}$
$Z_T = \sqrt{20^2 + (50 - 25)^2}$
$= 32 \ \Omega$

2. $I_T = {}^{120}/_{32}$ $I_T = 3.75$ A

3. $V_R = 3.75 \times 20$ $V_R = 75$ V

4. $V_{XL} = 3.75 \times 50$ $V_{XL} = 187.5$ V

5. $V_{XC} = 3.75 \times 25$ $V_{XC} = 93.75$ V

6. True Power $= 3.75^2 \times 20$
$P = 281.25$ W

7. Apparent Power $= 120 \times 3.75$
$S = 450$ V•A

8. Reactive Power $= 3.75^2 \times (50 - 25)$
$Q = 351.6$ VARS

9. Power Factor $= {}^{281.25}/_{450}$
$P.F. = 0.625$

RLC Parallel Circuits

AC circuits that contain resistance, inductance and capacitance connected in a parallel configuration are shown in illustration #41B. The RLC arrangement shown in illustration #41B is commonly found in industrial plants and employs the concept of improving the power factor by adding capacitors in parallel with inductive loads.

The advantage of connecting capacitors in parallel with inductive loads is that the line current decreases, and subsequently there will be a reduction in both line loss and line drop. Utility companies meter their commercial and industrial customers with either a kVA demand meter or with a combination kWHr meter and peak kVA meter. It often is advantageous for customers to install capacitors into their systems to reduce their total kVA demand. The following example will demonstrate the reduction that occurs in line values when capacitors are added.

RLC Parallel Circuits (cont'd)
RLC Parallel Circuit Example

Illustration #41A shows a system with an inductive load and no capacitors.

Illustration #41B shows a system with capacitors added to the system to improve the power factor.

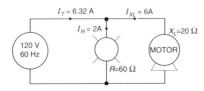

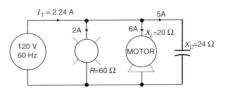

$$I_T = \sqrt{I_R^2 + I_{XL}^2}$$
$$I_T = \sqrt{2^2 + 6^2} = 6.32 \text{ A}$$
$$S = 120 \times 6.32 = 758 \text{ VA}$$
$$P = 2^2 \times 60 = 240 \text{ W}$$
$$\text{P.F.} = {}^{240}/_{758} = 0.317$$

$$I_T = \sqrt{I_R^2 + (I_{XL} - I_{XC})^2}$$
$$I_T = \sqrt{2^2 + (6-5)^2} = 2.24 \text{ A}$$
$$S = 120 \times 2.24 = 268 \text{ VA}$$
$$P = 2^2 \times 60 = 240 \text{ W}$$
$$\text{P.F.} = {}^{240}/_{268} = 0.896$$

Illustration #41A - No Capacitor

Illustration #41B - With Capacitor

Three-Phase Voltages - Wye

Most electrical power utilities generate three-phase voltages. There are significant advantages to the generation and utilization of three-phase power over single phase:

1. Three-phase equipment is smaller, lighter and more efficient than single phase equipment.

2. Three-phase rotating machinery operates more smoothly and quietly.

3. Three-phase rotating machinery is simpler in construction, more rugged and reliable.

4. Three-phase distribution systems require 75% of the copper wire to supply the same sized load as does a single phase system.

There are two basic wiring configurations for three-phase alternators and three phase loads. They may either be connected in wye or in delta.

Illustration #42 shows the displacement in electrical degrees of the voltages and currents for a wye connected circuit with a lagging power factor of .866 (balanced load).

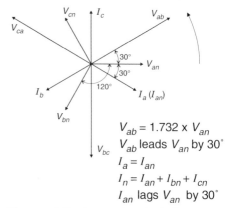

$$V_{ab} = 1.732 \times V_{an}$$
$$V_{ab} \text{ leads } V_{an} \text{ by } 30°$$
$$I_a = I_{an}$$
$$I_n = I_{an} + I_{bn} + I_{cn}$$
$$I_{an} \text{ lags } V_{an} \text{ by } 30°$$

Phase rotation is counter-clockwise.
Phase sequence is ABC.

Illustration #42 - Wye Connected Three-Phase System

Three-Phase Voltages - Wye/Delta

The following formulas and relationships apply to a **wye**-connected system or to a wye-connected load:

a. The phase voltage (V_p) is 1.732 times less than the line voltage (V_L).
b. The phase current (I_p) is equal to the line current (I_L).
c. The neutral current (I_N) is equal to the phasor sum of all the line currents.
d. The line voltages are displaced from each other by 120 electrical degrees.
e. The total apparent power (S) in a three phase circuit is equal to $V_L \times I_L \times 1.732$.

Note: The neutral wire must never be open circuited in a three-phase four-wire wye-connected system. Open circuiting the neutral will cause the magnitudes and angles of the phase voltages to shift. This can result in severe voltage unbalances and damage to electrical equipment. That is one reason why the neutral wire is not normally fused nor switched.

Illustration #43 shows the displacement, in electrical degrees of the voltages and currents in a delta-connected circuit with a lagging power factor of 0.866 (balanced load).

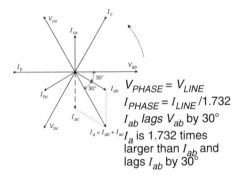

$$V_{PHASE} = V_{LINE}$$
$$I_{PHASE} = I_{LINE}/1.732$$
I_{ab} lags V_{ab} by 30°
I_a is 1.732 times larger than I_{ab} and lags I_{ab} by 30°

Phase rotation is counter-clockwise.
Phase sequence is ABC.

Illustration #43 - Delta Connected Three-Phase System

Three-Phase Voltages - Wye/Delta (cont'd)

For a **delta**-connected system, the following relationships and formulas apply:

a. The phase voltage (V_p) is equal to the line voltage (V_L).
b. The phase current (I_p) is 1.732 times less than the line current (I_L).
c. The line voltages are displaced from each other by 120 electrical degrees.
d. The total apparent power (S) in a three-phase system is equal to $V_L \times I_L \times 1.732$.

Three Phase Voltages - Sequences

Phase sequence refers to the order in which the three-phase voltages appear with respect to each other. There are only two phase sequences possible and both are shown in illustration #44. Note that all phasors are normally shown rotating in a counterclockwise direction.

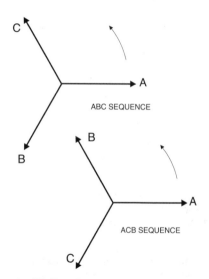

Illustration #44 - Phase Sequences

SECTION
TWO
ELECTRICAL CONDUCTORS

Conductor Materials

A perfect conductor would allow electric current to flow with no opposition. This perfect conductor would transport current without experiencing a rise in temperature or a drop in voltage. Scientists have been exploring the possibilities of just such a conductor called a superconductor that would operate at room temperature and offer no opposition to the flow of current. To date their endeavors have been met with limited success and superconductors for commercial use are as yet not available.

Materials that will readily allow current to flow may be used as electric conductors. The most popular and common materials used for conducting electrical currents are copper and aluminum. Table #5 lists the ten most common conducting materials and ranks them in order of their ability to conduct.

Each of the materials listed in table #5 has specialized uses and applications in conducting electrical current.

Common Conducting Materials	
Conductor Material	Relative Conductivity[1]
Silver	100%
Copper	95%
Gold	67%
Aluminum	58%
Tungsten	30%
Nickel	21%
Iron	14%
Constantan[2]	3.3%
Carbon	1.7%
Nichrome	1.5%

[1]All values are compared to silver.
[2]An alloy of copper and nickel.

Table #5 - Common Conducting Materials

Table #6 lists some common applications for each conducting material.

Applications of Conducting Materials	
Material	Application
Silver	Electrical contacts for relays and printed circuit boards
Copper	Electrical conductors for construction and equipment wiring
Gold	Plating of electrical conducting surfaces for sensitive electronic circuits
Aluminum	Long distance high voltage conductors
Tungsten	Filament material for lamps
Nickel	Plate material for batteries
Iron	Conducting path for magnetic circuits
Constantan	Thermocouples
Carbon	Resistors; Batteries
Nichrome	Heater elements

Table #6 - Conducting Material Applications

Copper

Copper is the most popular electrical conducting material and has many advantages over aluminum, its closest rival. As table #5 indicates, copper is only second to silver in its ability to conduct electrical current. Copper is also a readily available resource and remains relatively inexpensive in comparison to materials such as silver and gold. It exhibits low rates of oxidation and is easily formed or drawn into individual strands or conductors. Copper is very stable and does not expand or contract as readily as aluminum. From an installer's perspective, copper is easy to work with, handles well and leaves no messy discoloration on clothes, materials or hands.

When exposed to air copper will slowly oxidize; however, the oxide is a good conductor and will not significantly impair the ability of the metal to conduct electrical current.

Copper (cont'd)

Copper electrical wiring is usually annealed, that is, it is reheated and then slowly allowed to cool in order to render it tough and yet soft. Copper may also be hard drawn to obtain a stiffness and hardness that is useful for aerial suspension. Copper has the ability to be soldered allowing for connections in electronic printed circuit boards. Often copper busbars are tin, zinc, or silver plated to prevent oxidation and to prevent overheating wherever joints or connections are made.

Tinning of insulated conductors is justified where the immediate covering is not chemically compatible with copper or where the conductor is exposed to corrosive atmospheres and chemicals such as moist ammonia gas. Tin-plated copper conductors are recommended if water is constantly present at the conductor joints. Tin-plated copper conductors will have a slightly lower value of conductance than pure copper conductors. Table #7 provides a comparison of the characteristics of copper and aluminum.

Copper and Aluminum Comparison		
	Annealed Copper	Aluminum
Conductivity at 20°C; % of Copper	100	61.0
Resistivity ohms per circular mil-foot at 20°C	10.371	17.002
Resistivity Ω metres at 20°C	1.72×10^{-8}	2.83×10^{-8}
Specific gravity at 20°C	8.89	2.705
Melting point °C	1083	660
Melting point °F	1981	1220
Thermal expansion: inches per 1000 ft. of conductor per °F	.1128	.1536
Thermal expansion: mm per km per °C	16.92	23.04
Relative weight for equal direct-current resistance and length	1.0	0.50

Table #7 - Copper and Aluminum Comparison

Aluminum

The relative abundance of aluminum makes it a cost effective alternative to copper as a conductor for electrical wiring. It has some advantages over copper and is the second most commonly used material in present day electrical wiring. Aluminum has a good ampacity to weight ratio and therefore is often used in aerial wiring. Its light weight allows for long spans between supports and also its larger diameter for a given conductivity reduces corona (corona is the charging of air particles and occurs whenever a conductor is at a high voltage).

Note: Aluminum has lower conductivity properties than copper and will carry only 61% of the current of a similar sized copper conductor.

Aluminum has a few undesirable properties when used as a conductor and must therefore be handled and installed with care. An oxide film forms on the surface of aluminum just as it does with copper.

This oxide film acts as an insulator rather than as a conductor. The oxide film forms rapidly whenever fresh aluminum is exposed to oxygen. Under normal conditions it builds up to a thickness of one to two ten-millionths of an inch, (3 to 6×10^{-9} m). This oxide film can be both beneficial and problematic. It is beneficial in the sense that it acts as an insulating film or dielectric material and provides the aluminum with corrosion resistance. It is problematic at the termination points as oxide increases resistance, which causes the joint to heat up and the voltage to drop. Low resistance joints and connections must be made and maintained.

Another problem with aluminum is known as cold flow. All metals flow or accommodate when under mechanical pressure. Pure aluminum, being much softer than copper, will flow or accommodate under lower pressures. If approved connectors and techniques are not used in terminating aluminum, then a high-resistance heat-producing joint will result.

Aluminum (cont'd)

Aluminum when in contact with a dissimilar material such as copper and in the presence of an electrolyte such as water containing salt will undergo galvanic action. The unfortunate result of galvanic action is the dissolving of aluminum. To avoid this problem, bi-metallic connections must be made using approved connectors, and these connections must be kept dry.

Note: Different metals expand at different rates, and if copper and aluminum are terminated together, the connectors used must accommodate this expansion differential.

Handling and Terminating Aluminum

Aluminum is a very soft light-weight metal and allows for relative ease in installation. There are, however, some precautions that should be taken when working with aluminum:

1. Aluminum conductors can break or experience stress fractures after frequent bending. Try to reduce the amount of handling as much as is possible, and always avoid making sharp bends.

Note: A good basic rule to follow is to not bend the conductor beyond the point where deformation starts in the conductor or its insulation.

2. Aluminum, being a soft metal, will expand when its temperature increases. If long lengths of aluminum conductors are suspended or layed in trays or other raceways, it is necessary to account for a change in the conductors length with changes in temperature.

See table #7 for the thermal expansion of aluminum measured in inches per 1000 feet length per degree Fahrenheit increase or mm per 1000 metres per degree Celsius increase. The following example will help to explain the use of table #7.

Handling and Terminating Aluminum (cont'd)

Example#1:

A 492 foot (150 m) #3/0 aluminum cable is placed in a cable tray when the ambient temperature is 20°C (68°F). Calculate the change in the conductors length if the ambient temperature should rise to 40°C (104°F)

Example Solution (Imperial):

Lc = Ki x L x Tc

Lc = change in conductor length measured in inches

Ki = Imperial constant for aluminum from table #7

L = length of conductor in feet divided by 1000

Tc = change in ambient temperature in degrees F

Lc = 0.1536 x $^{492}/_{1000}$ x 36

Lc = 2.72 inches

Example Solution (Metric):

Lc = Km x L x Tc

Lc = change in conductor length measured in mm

Km = metric constant for aluminum from table #7

L = length of conductor in metres divided by 1000

Tc = change in ambient temperature in degrees C

Lc = 23.04 x $^{150}/_{1000}$ x 20

Lc = 69.12 mm

Connectors

Approved connector designs for aluminum have accommodated the need for increased contact areas and lower unit stresses. These terminals have adequate strength to ensure that the compression pressure of the aluminum is sufficient. They also produce a brushing action which destroys the oxide film, forms an intimate contact area, and results in a low-resistance connection.

Connectors (cont'd)

In general, aluminum connectors are most compatible with aluminum conductors.

Copper-bodied connectors of the mechanical or compression type are to be avoided since they are found to be unstable during heat-cycling tests; whereas, aluminum-bodied connectors are found to be compatible with both copper and aluminum.

Compression connectors as shown in illustration #45 have been found to be the most reliable. Failure of these connectors are rare and invariably attributable to factors other than their design.

Mechanical connectors such as solderless lugs or split bolts are alternative types of connectors. The advantage of mechanical connectors over compression connectors lies in their ability to be reversible; that is the conductor may be removed from the connector without damage and the connectors themselves may be salvaged (see illustration #46).

Illustration #45 - Compression Connectors

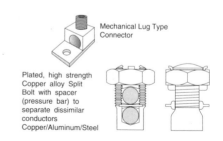

Mechanical Lug Type Connector

Plated, high strength Copper alloy Split Bolt with spacer (pressure bar) to separate dissimilar conductors Copper/Aluminum/Steel

Illustration #46 - Mechanical Connectors

Connectors (cont'd)

Care must be taken when removing insulation from aluminum conductors to ensure that the conductor does not get cut or nicked, as this will reduce the mechanical strength of the conductor.

In illustration #47 the preparation and connection of an aluminum conductor to a compression connector is shown. Illustration #48 indicates the preparation and connection of an aluminum conductor to a single hole solderless lug.

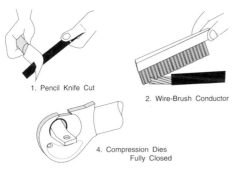

1. Pencil Knife Cut

2. Wire-Brush Conductor

4. Compression Dies Fully Closed

3. Conductor Inserted in CU/AL Compression Terminal

5. Completed Compression Terminal

Illustration #47- Compression Connector Preparation

Connectors (cont'd)

STEP 1. Carefully remove insulation without nicking conductor.

STEP 2. Wire brush conductor to remove any oxide.

STEP 3. Apply antioxidant to prevent formation of surface oxide.

STEP 4. Tighten mechanical connector securely.

Illustration #48 - Mechanical Connector Preparation

When aluminum lugs are connected to copper busbars and to steel or copper studs or bolts, care must be taken to prevent galvanic action and dissimilar thermal expansion. Illustration #49 shows the need to use a special kind of washer known as a *Belleville. Following this procedure will result in a trouble free connection.

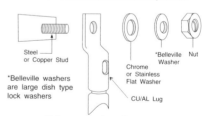

Bolting an Aluminum Compression Lug to a Steel or Copper Stud.

Illustration #49 - Aluminum to Steel or Copper Connection

Connectors (cont'd)

Belleville washers are not required where aluminum-to-aluminum contact is being made with an aluminum bolt.

If a connector or lug is bolted to unplated aluminum busbars, the busbar material should first be lightly wire brushed and then an approved aluminum joint compound applied over the contact area. See illustration #50 for typical compounds.

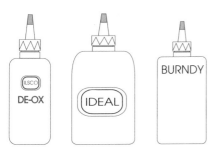

Illustration #50 - Aluminum Connector Compounds

Code Requirements

When terminating solid aluminum conductors at receptacles and switches, both the Canadian Electrical Code and the National Electrical Code require that certain procedures be followed. The steps are shown in illustration #51 and described below:

1. Remove the insulation with an approved wire stripper. If a knife is used, remove the insulation using a pencilling cut. Avoid cutting a ring around the insulation and then pulling the insulation off. Ringing the insulation may result in nicking the wire, thus weakening its mechanical strength.

2. Connect the conductor only to UL or CSA listed devices that are marked CO/ALR. Never connect to push-in, back wired or CU/AL marked devices.

3. Loop the wire three quarters of the way around the screw head and in the direction of tightening rotation. Only one conductor is to be connected to any one screw.

Code Requirements (cont'd)

4. Tighten the screw head until the conductor is firmly in contact with the underside of the screw head and the contact plate on the wiring device.

5. Tighten the screw an additional half turn to ensure a firm connection.

6. To decrease the likelihood of the wires becoming loose when the device is inserted into the device box, make a U shaped loop in the wires behind the wiring device.

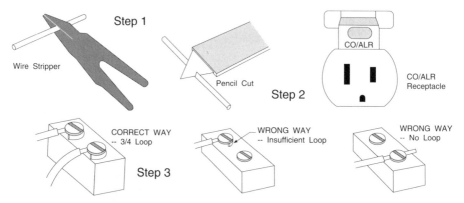

Illustration #51 - Code Requirements for Connecting Aluminum Conductors

Conductor Insulation

Under normal and expected conditions electrical current should only flow to the intended load. If current flows to ground or between load conductors, a condition known as insulation breakdown has occured. The conductor's insulation is designed to prevent the flow of current to any point other than the intended load. All electrical circuit conductors must be insulated. Even if the conductors are bare, they must be spaced far enough apart from each other and the ground to ensure that the air acts as an adequate insulator.

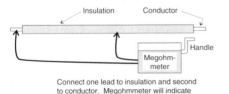

Connect one lead to insulation and second to conductor. Megohmmeter will indicate very high resistance if insulation is good.

Illustration #52 - Megohmmeter Insulation Resistance Test

Insulated conductors are coated in materials that exhibit high insulation resistance and high dielectric strength. Insulation resistance refers to the resistance to leakage current either through or over the surface of the insulation material. Insulation resistance can be measured and tested by using a megohmmeter as shown in illustration #52, or by some other suitable testing means.

Dielectric strength is the ability of a material to withstand electrical stress and is expressed in kilovolts per millimeter (see table #4A, page 42, Section One).

The dielectric strength of the insulation must be greater than the electrical pressure, measured in volts, that the conductor will be operating at. A cable that is connected to a 5 kV source must be able to withstand a continuous pressure of 5000 V and not fail throughout its rated life.

Conductor Insulation (cont'd)

Manufacturers of cables provide a voltage rating for each cable or conductor. Common values are 600 V, 1000 V, 5000 V, 15,000 V and 25,000 V.

Conductors that are operating at high voltages (greater than 750 V) have an insulation level rating of either 100%, 133% or 173%. This rating relates to the clearing time of the overcurrent device used to protect the cables. See Section Nine, page 418 for further information.

Insulation Materials

There are a wide variety of conductor insulations available, with the most common ones being thermoplastic, nylon, neoprene, varnish, glass coated over a thermoplastic jacket, crosslinked polyethylene, ethylene propylene rubber (EPR), paper impregnated with oil, and mineral insulated (magnesium oxide). Each of these materials has its own advantages, disadvantages and applications.

Thermoplastic: Polyethylene, polyvinyl chloride, and nylon are all classified as thermoplastics. Thermoplastic is flexible, has good insulation resistance and is moisture resistant. It is sensitive to high temperatures, and at very low temperatures becomes brittle and may crack.

Thermoplastics are limited to 5 kV or less as the material will experience heating even though no current is flowing. This heating is caused by dielectric stressing.

Common Canadian manufacturers' designations for thermoplastic insulation are TW, TW75, TWU, TWU75, T90 NYLON and THHN. Table #8 contains a list of each of these insulations. Table #9 has USA trade names and type letters assigned to insulated conductors.

Nylon: Nylon is rigid, extremely tough and resistant to abrasion and is a good material for an oversheath or jacket.

Thermoplastic Insulation Designations (Canada)			
Common designation and description	**Minimum installation temp.**	**Maximum operating temp.**	**Applications**
TW: T (thermoplastic) W (wet)	-40°C	60°C	For exposed wiring and in raceways
TW75: T (thermoplastic) W (wet) 75 (temperature rating)	-10°C	75°C	For exposed wiring and in raceways
TWU: T (thermoplastic) W (wet) U (underground)	-40°C	60°C	For direct earth burial, for open wiring exposed to the weather, for service entrance
TWU75: T (thermoplastic) W (wet) U (underground) 75 (temperature rating)	-10°C	75°C	For direct earth burial, for open wiring exposed to the weather for service entrance
THHN T90 NYLON: T (thermoplastic) 90 (temp rating) NYLON (PVC insulation with nylon covering)	-10°C	90°C 60°C if exposed to oil	For general purpose building wire

Table #8 -Thermoplastic Insulations (Canada)

Insulation Materials (cont'd)

Thermoplastic Insulation Designations (USA)			
Trade Name	**Type Letter**	**Maximum Operating Temperature**	**Applications**
Moisture resistant thermoplastic	TW	60°C	Dry and wet locations
Moisture and heat resistant thermoplastic	THHW	75°C	Dry and wet locations
Heat resistant thermoplastic	THHN	90°C	Dry and damp locations

Table #9 - Thermoplastic Insulations (USA)

Neoprene: Neoprene was the first commercial synthetic rubber. Its major use is as a tough and flexible sheathing material and is commonly used in flexible cords and portable lines. It exhibits very good abrasion and tear resistance, together with good resistance to swelling and to chemical attack. It is impervious to oil and makes an excellent insulation material for power cords used in the petrochemical industry.

Neoprene does not normally support combustion. The abbreviation for heat-proof neoprene is HPN.

Varnish: Varnish is usually baked on to copper conductors that are used in motor and transformer windings. Copper conductors are dipped into a varnish bath and then allowed to bake dry. The temperature rating of varnish baked-dry insulated conductors is in excess of 200°C.

Insulation Materials (cont'd)

Glass Coated Over a Thermoplastic Jacket: A highly heat resistant insulation rated 125°C. Commonly used as equipment wiring for heat producing devices such as recessed luminaires. A common wire designation is GTF (Glass and Thermoplastic for Fixtures). This insulation is to be used for equipment wiring in dry locations only.

Crosslinked Polyethylene (XLPE): Polyethylene has excellent electrical properties as an insulator but is limited to a 70°C operating temperature. By incorporating an agent such as peroxide, or by subjecting the polyethylene to radiation, the working temperature of the insulation is increased to 90°C. This crosslinking with either a peroxide or by radiation is what enhances the use of polyethylene and has given rise to its insulation popularity for general building wire. Crosslinked polyethylene is classified as a thermoset polymer and, therefore, exhibits excellent mechanical and thermal properties.

Crosslinked polyethylene does however exhibit memory. It is somewhat stiff. It does not have good resistance to oils and certain gases. Canadian manufacturer's designations are RW-90 XLPE, RWU-90 XLPE and trade names like Exelene, Vulkene, Thermalene, etc. USA designations are XHHW and XHHW-2.

Ethylene Propylene Rubber (EPR): EPR exhibits similar electrical properties to that of XLPE but has the added enhancement of being more flexible and softer. EPR is classified as a thermoset polymer and is crosslinked with more compounds than XLPE. It has a working temperature of 90°C. Common designations for this insulation are RW90 EP and RWU90 EP.

Because of its superior performance, flexibility and suitability for continuous operation at 90°C, it has become a popular choice for use on portable cables and is commonly used in the mining industry.

Insulation Materials (cont'd)

Paper Impregnated With Oil: Dry paper acts as a very good insulator. To ensure that it remains dry it is impregnated with an oil such as a medium viscosity polybutane-based compound containing no added oxidation inhibitors. Paper construction is generally of 2-ply form, but some 3-ply is used at higher voltages. Each ply of paper is approximately 3 to 7 mils (.065 to .190 mm) thick. Insulation thicknesses vary from 24 to 1200 mils (0.6 to 30 mm).

The thickness of the paper construction insulation depends on the impressed voltage and the electrical stress. As the insulation becomes thicker bending problems become greater. The paper insulation must be protected against mechanical damage and therefore, is enclosed in a metal sheath. The sheath is usually made of lead or a lead alloy such as tin cadmium lead, tin antimony lead, arsenical lead or copper-bearing lead.

This type of insulation is used on cables that operate at high voltages, usually 69 kV or higher. Illustration #53 shows a paper-insulated lead covered cable (PILC).

Note: Because of the lead the handling of PILC can be harmful and therefore the cable is rarely used in new installations.

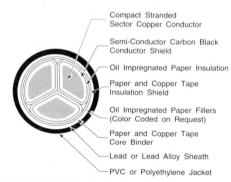

Compact Stranded Sector Copper Conductor

Semi-Conductor Carbon Black Conductor Shield

Oil Impregnated Paper Insulation

Paper and Copper Tape Insulation Shield

Oil Impregnated Paper Fillers (Color Coded on Request)

Paper and Copper Tape Core Binder

Lead or Lead Alloy Sheath

PVC or Polyethylene Jacket

Illustration #53 - Paper Insulated Lead Covered Cable

Insulation Materials (cont'd)

Mineral Insulated: This type of insulation had its beginnings in 1896 when a Swiss engineer proposed a cable be constructed from inorganic materials and be completely enclosed in a metal sheath. In 1936 a British Company, Pyrotenax Limited, began producing mineral insulated cable.

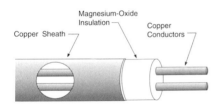

Illustration #54 - Mineral Insulated Cable

Current carrying copper conductors are embedded in a highly compacted mineral insulation (magnesium-oxide) and then enclosed in a copper or stainless steel sheath. To protect the copper sheath from corrosive environments a thermoplastic sheath may be applied, or a stainless steel sheathing material may be used. Illustration #54 shows the construction of a typical mineral insulated (MI) cable.

This type of cable performs extremely well under both high temperatures and when subjected to severe mechanical abuse such as bending, twisting and impact.

Copper sheath cables are resistant to most organic chemicals and can operate in most industrial environments. MI cable also has the added characteristic of being able to withstand radiation.

Insulation Materials (cont'd)

Mineral Insulated (cont'd): Magnesium oxide has been found to be chemically and physically stable, has a high melting temperature, high electrical resistivity combined with high thermal conductivity, is non-toxic and is readily available.

MI cable is commonly used for critical circuits such as wiring for emergency systems, in petrochemical plants, in areas that experience high temperatures such as a smelter, in mines, on ships, and on the space shuttle's launch pad.

Fire Rating

Both Canada and the USA require that some locations be wired with conductors or cables that have a fire rating. In the USA, insulation and outer coverings that meet the requirements of flame-retardant limited smoke are to be designated limited smoke with the suffix /LS after the Code designation.

In Canada the 1994 edition of the Electrical Code requires electrical cables and wires that are to be installed in buildings to meet flame spread requirements that are mandated by the National Building Code (NBC) or by local legislation.

Cables and conductors may be marked either FT1, FT2, FT4, FT5 or FT6. FT1 rated wires and cables are suitable for installation in buildings of combustible construction, and FT4 wires and cables are suitable for installation in noncombustible and combustible construction. Wires and cables with combustible outer jackets or sheaths that do not meet the FT1 or FT4 classification should be located in noncombustible raceways, masonry walls or concrete slabs.

Conductor Shapes

Electrical conductors are available in various shapes including; round, oval, "D" shape, rectangular, wedge, and can be formed into solid, hollow or stranded conductors.

Note: North American cable size standards, which are followed in this book, are different than European and British standards.

Cable manufacturers employ different conductor shapes depending upon the voltage and application of service. Round conductors are the most common for sizes up to #6 AWG or 16 mm^2 and for higher voltages where it is important to distribute the electrical stress equally. Illustration #55 shows sector shaped conductors. These shapes, both the "D" shape and the oval shape, are used to help keep the cable dimensions at a minimum. A "D" shape is frequently used for 2-core cables and an oval construction for 33 kV cables.

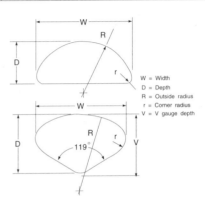

W = Width
D = Depth
R = Outside radius
r = Corner radius
V = V gauge depth

Illustration #55 - Multicore Cable Sector Shapes

Rectangular conductors are usually referred to as busbars and are found in manufactured switchgear or in busducts. They are composed of either solid copper or aluminum. The ampacity of these conductors depend upon a number of factors that will be discussed later in this section.

Conductor Shapes (cont'd)

Hollow copper and hollow aluminum conductors are usually round and used in high voltage distribution networks. The reason for using hollow versus solid conductors is to reduce a phenomena called skin effect as well as corona loss. Electrical current tends to travel on the outer surface or the skin of a conductor. Outdoor switch yards make use of hollow conductors resulting in lower costs and lower conductor weight. Skin effect becomes more noticeable for AC cables larger than 300 kcmil at 60 Hz.

Stranded conductors are used in cables to increase their flexibility. For general construction wiring, solid conductors are permitted up to size #10 AWG or 6 mm², after which stranded conductors are required. A conductor of a particular size can be stranded with a few heavy strands of wire or with many light strands. Regardless of the number of strands, the total area remains the same.

When using heavy strands of wire as shown in illustration #56A, the spaces between the conductors is greater, and this may result in a larger cable diameter.

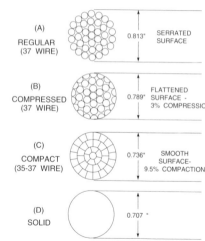

(A) REGULAR (37 WIRE) — 0.813" SERRATED SURFACE

(B) COMPRESSED (37 WIRE) — 0.789" FLATTENED SURFACE - 3% COMPRESSION

(C) COMPACT (35-37 WIRE) — 0.736" SMOOTH SURFACE - 9.5% COMPACTION

(D) SOLID — 0.707"

Class B Stranded Conductors 500 kcmil (MCM)
Copper or Aluminum as compared to a Solid Conductor

Illustration #56 - Conductor Configuration and Dimensions

Conductor Shapes (cont'd)

A #3/0 AWG or 85 mm² conductor could be solid (1 strand, used for a lightning rod), 19 conductors concentrically arranged (eg: for building cables), or 5340 conductors (eg: for very flexible welding cables). All three configurations result in the same amount of conducting area. Illustration #56 shows the different conductor arrangements available for a class B stranded 500 kcmil (MCM) conductor.

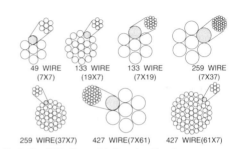

Illustration #57B - Concentric Rope-Lay Strands

Concentric stranded cables or conductors are formed by having identically sized individual wires or strands laid together in a circular configuration and then layered concentrically as shown in illustration #57A and #57B.

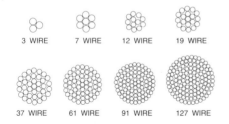

Illustration #57A- Concentric Layer Strands

Conductor Shapes (cont'd)

Each layer forms a definite mathematical pattern. Illustration #58 indicates the emerging pattern as layers are added to the cable. The first layer contains one strand, the second layer contains six strands, the third layer contains twelve strands, the fourth layer contains eighteen strands, etc.

North American practice classifies various concentric stranded formations by the total number of wires or strands in the conductor for any given conductor size. Table #10 lists these stranding classifications.

Illustrations #59A and #59B show some of the sizes, shapes, materials and applications of conductors.

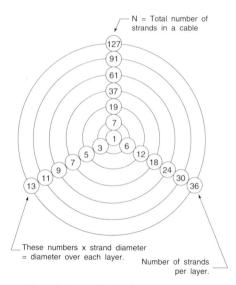

N = Total number of strands in a cable

These numbers x strand diameter = diameter over each layer.

Number of strands per layer.

Illustration #58- Layered Cable Pattern

Conductor Wire Strand Classifications	
Class AA Stranding	usually specified for bare conductors on overhead lines
Class A Stranding	usually specified for weatherproof (ie. weather resistant covering) conductors for overhead lines or where greater flexibility than Class AA is needed
Class B Stranding	usually specified for insulated overhead and underground conductors (cables); classified as "standard" building wire and cable
Class C Stranding	classified as "flexible" building wire and cable
Class D Stranding	classified as "extra flexible" building wire and cable
Class G Stranding	for portable cables
Class H Stranding	for highly flexible cords and cables
Class I Stranding	for apparatus cables and motor leads
Class K Stranding	for special flexible portable cords
Class M Stranding	for welding cables etc.

Table #10 - Wire Stranding Classifications

Conductor Data			
Material	**Type of Configuration**	**Shape**	**Sizes and Applications**
Copper	Solid, round		- Bare wire in sizes No. 4/0 to 45 AWG - Covered non-insulated sizes No. 4/0 to 14 AWG - Insulated in sizes No. 6 to 14 AWG for power work
	Solid, rectangular		Busbars. Available in a variety of sizes eg. 1/4" x 2"
	Solid, tubes		Busbars. Available in a variety of sizes
	Standard, concentric-stranded		Bare, covered and insulated in sizes 5000 kcmil (MCM) to No. 20 AWG
	Standard, rope-stranded	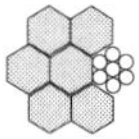	Rubber-sheathed cords and cables in sizes 5000 kcmil (MCM) to No. 8 AWG

Illustration #59A - Conductor Data

Conductor Data			
Material	**Type of Configuration**	**Shape**	**Sizes and Applications**
Copper	Compact-stranded, sector		Multiconductor paper insulated cables, both solid and oil-filled types in sizes 1,000 kcmil (MCM) to No. 1/0 AWG
	Hollow-core, stranded		Oil-filled paper-insulated single conductor cables in sizes 2,000 kcmil (MCM) to No. 2/0 AWG
Aluminum	Solid, round Solid, tubes		Bare wire in sizes No. 4/0 to 29 AWG. Covered non-insulated in sizes No. 4/0 to 12 AWG. Insulated in sizes No. 6 to 12 AWG for power work.
	Standard, Concentric-Stranded		Bare or weatherproof covered in sizes 1,600 kcmil (MCM) to No.6 AWG
	Annular aluminum, stranded with steel core		Bare in sizes 1,600 kcmil (MCM) to No.4 AWG. Weatherproof covered in sizes 4/0 to 8 AWG

Conductor Sizes

Materials that freely permit the motion of electrons are classified as conductors. The ability of a conductor to allow electrons to move freely depends on its atomic structure (type of material) and its physical size (cross-sectional area). In order to compare the size of one conductor to another, a standard unit has been established.

In North America this unit is based on a linear length called a mil (0.001 of an inch). Wire gage numbers are constructed using this base unit. The American Wire Gage (AWG), formerly called the Brown and Sharpe Gage (B&S), is used in both Canada and the U.S.A.

Round and rectangular conductors are measured and gaged based on the mil. A rectangular conductor is measured in square mils. A square mil is an area of a square the sides of which are 1 mil, as shown in illustration #60.

Square or Rectangular Busbar Area

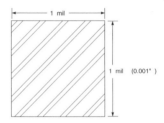

Illustration #60 - Busbar Expressed in Square Mils

To obtain the cross-sectional area of a square conductor, simply square its side.

The cross-sectional area in square mils of a rectangular conductor is found by multiplying the length of one side by the length of the other side, each length expressed in mils. Example #1 illustrates the calculations involved in determining the size of a busbar expressed in square mils.

Rectangular Conductor Area

Example #1:

Determine the cross-sectional area of a rectangular copper busbar that is 1/4 inch thick and 2 inches wide.

Step 1. Convert the Imperial units to mils:
(1 inch = 1000 mils)
1/4 in. = 250 mils, 2 in. = 2000 mils

Step 2. Cross-sectional area
= thickness x width:
250 x 2000 = 500,000 square mils

Round Conductor Area

The circular mil (cmil) is the standard unit of wire and cross-sectional area used in American wire tables. The circular mil (cmil) is based on a conductor's diameter being measured in mils (1/1000 of an inch). Because the diameters of round conductors may be only a small fraction of an inch, it is convenient to express these diameters in mils to avoid fractions.

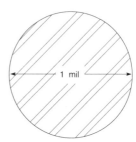

Illustration #61 - Circular Mil

A circular mil (cmil) is the area of a conductor having a diameter of 1 mil as shown in illustration #61. The area of a round conductor may be expressed in circular mils by squaring its diameter, expressed in mils. Example #2 shows how the area of a round conductor may be calculated, either in circular mils or square inches.

Round Conductor Area (cont'd)
Example #2:

Calculate the area of a round conductor that has a diameter of 1/2 inch. Express the answer in circular mils and in square inches.

Step 1. Change the diameter in inches to mils
1/2 inch = 500 mils

Step 2. Area = diameter^2
Area = 500 x 500
Area = 250,000 cmil

Step 3. Area expressed in square inches is based on the formula
$A = \pi \ (D/2)^2$
$A = 3.14 \ (0.5/2)^2$
$A = 0.19625$ square inches

Round vs Rectangular Comparison

The area of a square conductor may be compared to that of a round conductor.

As shown in illustrations #60 and #61, a square conductor having each side the same length as the diameter of a round conductor will have a greater area. This difference can be quickly determined by calculating the area of both a round conductor with a diameter of 1 mil, and a square conductor with a length of 1 mil.

Example #3:

The area of a square conductor having each side 1 mil in length:
$A = L^2$
$A = 1 \times 1$
$A = 1$ sq. mil

The area of a round conductor with a diameter of 1 mil, expressed in square mils:
$A = \pi \ (D/2)^2$
$A = 3.14 \ (1/2)^2$
$A = .7854$ sq. mils

Round vs Rectangular Comparison (cont'd)

Note: The area of a circular mil is equal to 0.7854 of a square mil. This relationship will be handy to remember whenever a conversion to either unit needs to be done.

Example #4 shows how to calculate the area of a square conductor in circular mils.

Example #4:

Calculate the area of a copper busbar that has a dimension of 1/2 in. x 2 in. Express the value of the area in equivalent circular mils.

Step 1. Determine the area of the busbar in square mils.

$$A = 500 \times 2000$$
$$A = 1,000,000 \text{ sq. mils}$$

Step 2. Convert the area from square mils to circular mils.

0.7854 sq. mils = 1.0 cmil
Area in cmil = 1,000,000/0.7854
A = 1,273,237 cmil

Note: As seen in example #4, the area of a square conductor is converted to circular mils by dividing the value of square mils by 0.7854. Conversely, the area of a round conductor expressed in circular mils can be converted to square mils by taking the circular mils and multiplying by 0.7854.

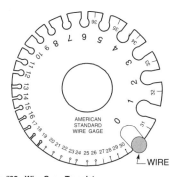

Illustration #62 - Wire Gage Template

Size of Round Conductors

The American Wire Gage (AWG) expresses the sizes of wires in gage numbers. The wire gage template may be used to measure the size of a conductor expressed in AWG numbers. See illustration #62.

Note: A micrometer can also be used to determine the size of a circular conductor by obtaining the diameter and then squaring the value as shown in example #2, page 80. The calculated value of the conductor expressed in circular mils can then be referred to a table to obtain the nearest wire gage. Table #11 (solid copper), table #12 (solid aluminum), table #13A and #13B (stranded copper), and table #14A and #14B (stranded aluminum) list the AWG numbers, their equivalent values in circular mils and other technical information.

There is a quick and easy method to approximate the circular mil area and gage number of conductors. A #10 AWG copper conductor has an area of approximately 10,000 cmil and a diameter of 100 mils. The conductor has approximately 1 ohm of DC resistance for each 1000 feet of length at 68°F (20°C). All other sizes can be determined from the #10 conductor as follows:

- One increase in the gage number represents a decrease of 26% in cross-sectional area.

- One decrease in the gage number represents an increase of 26% in cross-sectional area.

Round Conductors (cont'd)

- On the basis that a decrease of one gage number results in a conductor with a 26% greater cross-sectional area, a decrease of 3 gage numbers will produce a conductor with a cross sectional area twice as large (1.26 x 1.26 x 1.26 = 2). The converse is also true, an increase of three gage numbers will produce a conductor with half the cross-sectional area.

- When the cross-sectional area is halved, the DC resistance is doubled. When the cross sectional area is doubled, the DC resistance is halved.

- Every 10 gage sizes up or down from a reference value will result in a change of ten times or one tenth of the original value.

Example #5:

Using the known values of a #10 AWG copper conductor (area ≈ 10,000 cmil, Rdc ≈ 1.0 Ω) determine the approximate values of the following gage numbers:

a. #14 AWG
b. #6 AWG
c. #1/0 AWG

a. A #14 AWG is 4 gage numbers smaller than a #10 AWG. It has a cross sectional area of 10,000 / (1.26 x 1.26 x 1.26 x 1.26) ≈ 4000 cmil and a resistance of 1.00 x 1.26 x 1.26 x 1.26 x 1.26 ≈ 2.52 Ω.

b. A #6 AWG is 4 gage numbers larger than a #10 AWG. It has a cross sectional area of 10,000 x 1.26 x 1.26 x 1.26 x 1.26 ≈ 25,000 cmil and a resistance of 1.00 / (1.26 x 1.26 x 1.26 x 1.26) ≈ 0.397 Ω

c. A 1/0 AWG conductor is exactly 10 gage numbers larger than a #10 AWG conductor. The area is 10,000 x 10 ≈ 100,000 cmil and the resistance is 1.0/10 ≈ 0.1 Ω

Circular Mils/ Wire Gage - Solid Copper									
Size of Wire		Diameter		Area		Approx. net Cable Weight		Nominal DC Resistance (25°C)	
AWG	Circ. Mils	mm	in	mm²	sq.in.	kg per 1000 m	lbs per 1000 ft	Ohms/ 1000 m	Ohms/ 1000 ft
30	100	0.254	.0100	0.051	.000078	0.450	.303	347	106.0
29	128	0.287	.0113	0.065	.000100	0.575	.387	272	82.8
28	159	0.320	.0126	0.080	.000125	0.715	.481	219	66.6
27	202	0.361	.0142	0.102	.000158	0.908	.610	172	52.4
26	253	0.404	.0159	0.128	.000199	1.14	.765	137	41.8
25	320	0.455	.0179	0.162	.000252	1.44	.970	108	33.0
24	404	0.511	.0201	0.205	.000317	1.82	1.22	85.9	26.2
23	511	0.574	.0226	0.259	.000401	2.30	1.55	67.9	20.7
22	640	0.643	.0253	0.324	.000503	2.88	1.94	54.2	16.5
21	812	0.724	.0285	0.412	.000638	3.66	2.46	42.7	13.1
20	1020	0.813	.0320	0.519	.000804	4.61	3.10	33.9	10.3
19	1290	0.912	.0359	0.653	.001010	5.81	3.90	26.9	8.21
18	1620	1.024	.0403	0.823	.001280	7.32	4.92	21.4	6.51
17	2050	1.151	.0453	1.040	.001610	9.24	6.21	16.9	5.15
16	2580	1.290	.0508	1.310	.002030	11.6	7.81	13.4	4.10
15	3260	1.450	.0571	1.650	.002560	14.7	9.87	10.6	3.24
14	4110	1.628	.0641	2.080	.003230	18.5	12.4	8.44	2.57
13	5180	1.829	.0720	2.630	.004070	23.4	15.7	6.69	2.04
12	6530	2.052	.0808	3.310	.005130	29.4	19.8	5.31	1.62
11	8230	2.340	.0907	4.170	.006460	37.1	24.9	4.22	1.29
10	10380	2.588	.1019	5.260	.008155	46.8	31.4	3.34	1.02
9	13090	2.906	.1144	6.630	.010280	59.0	39.6	2.65	0.808
8	16510	3.264	.1285	8.370	.012970	74.4	50.0	2.10	0.640
7	20820	3.665	.1443	10.550	.016350	93.8	63.0	1.67	0.508
6	26240	4.115	.1620	13.300	.020610	118.2	79.4	1.32	0.403
5	33090	4.620	.1819	16.770	.025990	149.0	100.2	1.05	0.320
4	41740	5.189	.2043	21.150	.032780	188.0	126.3	0.831	0.253
3	52620	5.827	.2294	26.670	.041330	237.1	159.3	0.659	0.201
2	66360	6.543	.2576	33.620	.052120	298.9	200.9	0.523	0.159
1	83690	7.348	.2893	42.410	.065730	377.0	253.3	0.415	0.126
1/0	105600	8.252	.3249	53.490	.082910	475.5	319.5	0.329	0.1002
2/0	133100	9.266	.3648	67.430	.104500	599.5	402.8	0.261	0.0795
3/0	167800	10.40	.4096	85.010	.131800	755.8	507.8	0.207	0.0630
4/0	211600	11.68	.4600	107.220	.166200	953.2	640.5	0.164	0.0500

Table #11 - Circular Mils/Wire Gage Solid Copper

Circular Mils/Wire Gage - Solid Aluminum									
Size of Wire		Diameter		Area		Approx. net Cable Weight		Nominal DC Resistance (25°C)	
AWG	Circ. Mils	mm	in	mm^2	sq.in.	kg per 1000 m	lbs per 1000 ft	Ohms/ 1000 m	Ohms/ 1000 ft
20	1020	0.813	.0320	0.519	.000804	1.40	.942	55.6	16.9
19	1290	0.912	.0359	0.653	.001010	1.77	1.19	44.2	13.5
18	1620	1.024	.0403	0.823	.001280	2.22	1.49	35.0	10.7
17	2050	1.151	.0453	1.040	.001610	2.81	1.89	27.7	8.45
16	2580	1.290	.0508	1.310	.002030	3.53	2.38	22.1	6.72
15	3260	1.450	.0571	1.650	.002560	4.47	3.00	17.5	5.32
14	4110	1.628	.0641	2.080	.003230	5.63	3.78	13.8	4.22
13	5180	1.829	.0720	2.630	.004070	7.10	4.77	11.0	3.35
12	6530	2.052	.0808	3.310	.005130	8.94	6.01	8.72	2.66
11	8230	2.304	.0907	4.170	.006460	11.3	7.57	6.92	2.11
10	10380	2.588	.1019	5.260	.008155	14.2	9.56	5.48	1.67
9	13090	2.906	.1144	6.630	.010280	17.9	12.0	4.35	1.33
8	16510	3.264	.1285	8.370	.012970	22.6	15.2	3.45	1.05
7	20820	3.665	.1443	10.550	.016350	28.5	19.2	2.73	0.833
6	26240	4.115	.1620	13.300	.020610	35.9	24.1	2.17	0.661
5	33090	4.620	.1819	16.770	.025990	45.3	30.5	1.72	0.524
4	41740	5.189	.2043	21.150	.032780	57.2	38.4	1.36	0.416
3	52620	5.827	.2294	26.670	.041330	72.1	48.4	1.08	0.330
2	66360	6.543	.2576	33.620	.052120	90.9	61.1	0.858	0.261
1	83690	7.348	.2893	42.410	.065730	114.6	77.0	0.680	0.207
1/0	105600	8.252	.3249	53.490	.082910	144.6	97.2	0.539	0.164
2/0	133100	9.266	.3648	67.430	.104500	182.3	122.5	0.428	0.130
3/0	167800	10.40	.4096	85.010	.131800	229.8	154.4	0.339	0.103
4/0	211600	11.68	.4600	107.220	.166200	289.8	194.7	0.269	0.082

Table #12 - Circular Mils/Wire Gage Solid Aluminum

Circular Mils/Wire Gage - Stranded Copper										
Conductor Size		No.of Wires	Wire Diameter (Individual Strands)		Class "B" Standard		Compressed Round		Compact Round	
AWG	Circ. Mils		mm	in	mm	in	mm	in	mm	in
20	1020	7	0.31	.0121	.91	.036				
18	1620	7	0.39	.0152	1.17	.046				
16	2580	7	0.49	.0192	1.47	.058				
14	4110	7	0.61	.0242	1.85	.073	1.80	.071		
12	6530	7	0.77	.0305	2.34	.092	2.26	.089		
10	10380	7	0.98	.0385	2.95	.116	2.87	.113		
8	16510	7	1.23	.0486	3.71	.146	3.61	.142	3.40	.134
6	26240	7	1.55	.0612	4.67	.184	4.55	.179	4.29	.169
4	41740	7	1.96	.0772	5.89	.232	5.72	.225	5.41	.213
3	52620	7	2.20	.0867	6.60	.260	6.40	.252	6.05	.238
2	66360	7	2.47	.0974	7.42	.292	7.19	.283	6.81	.268
1	83690	19(18)*	1.69	.664	8.43	.332	8.18	.322	7.59	.299
1/0	105600	19(18)*	1.89	.0745	9.47	.373	9.19	.362	8.53	.336
2/0	133100	19(18)*	2.13	.0837	10.64	.419	10.31	.406	9.55	.376
3/0	167800	19(18)*	2.39	.0940	11.94	.470	11.58	.456	10.70	.423
4/0	211600	19(18)*	2.68	.1055	13.41	.528	13.00	.512	12.10	.475
250	MCM	37(35)*	2.09	.0822	14.61	.575	14.17	.558	13.20	.520
300	/kcmil	37(35)*	2.29	.0900	16.00	.630	15.52	.611	14.50	.570
350		37(35)*	2.47	.0973	17.30	.681	16.79	.661	15.60	.616
400		37(35)*	2.64	.1040	18.49	.728	17.93	.706	16.70	.659
500		37(35)*	2.95	.1162	20.65	.813	20.04	.789	18.70	.736
600		61(58)*	2.52	.0992	22.68	.893	22.00	.866	20.70	.813
750		61(58)*	2.82	.1109	25.35	.998	24.59	.968	23.10	.908
1000		61(58)*	3.25	.1280	29.26	1.152	28.37	1.117	26.90	1.060
1250		91	2.98	.1172	32.74	1.289	31.75	1.250		
1500		91	3.26	.1284	35.86	1.412	34.80	1.370		
1750		127	2.98	.1174	38.76	1.526	37.59	1.480		
2000		127	3.19	.1255	41.45	1.632	40.21	1.583		

NOTE:* Reduced number of wires for compact stranding are shown in parentheses

Table #13A - Circular Mils/Wire Gage Stranded Copper

Circular Mils /Wire Gage - Stranded Copper						
Conductor Size	Area		Approx. net Cable Weight		Nominal DC Resistance (25°C)	
AWG	mm^2	sq.in.	kg per 1000 m	lbs per 1000 ft	Ohms/1000 m	Ohms/1000 ft
20	0.517	.00080	4.69	3.15	34.6	10.5
18	0.821	.00128	7.46	5.02	21.8	6.64
16	1.31	.00203	11.90	7.97	13.7	4.18
14	2.08	.00323	18.90	12.7	8.61	2.63
12	3.31	.00513	30.00	20.2	5.42	1.65
10	5.26	.00816	47.70	32.1	3.41	1.04
8	8.37	.01297	75.90	51.0	2.14	.653
6	13.30	.02061	121	81.1	1.35	.411
4	21.15	.03278	192	129	0.848	.258
3	26.66	.04133	242	163	0.673	.205
2	33.63	.05212	305	205	0.533	.163
1	42.41	.06573	385	258	0.423	.129
1/0	53.51	.08291	485	326	0.335	.102
2/0	67.44	.1045	611	411	0.266	.0811
3/0	85.03	.1318	771	518	0.211	.0643
4/0	107.22	.1662	972	653	0.167	.0510
250	128.68	.1963	1149	772	0.142	.0432
300	152.01	.2356	1378	926	0.118	.0360
350	177.35	.2749	1609	1081	0.101	.0308
400	202.68	.3142	1838	1235	0.0885	.0270
500	253.35	.3927	2298	1544	0.0708	.0216
600	304.02	.4712	2758	1853	0.0590	.0180
750	380.03	.5890	3447	2316	0.0472	.0144
1000	506.71	.7854	4595	3088	0.0354	.0108
1250	633.38	.9817	5743	3859	0.0283	.00863
1500	760.06	1.1780	6892	4631	0.0236	.00719
1750	866.74	1.3740	8041	5403	0.0202	.00616
2000	1013.42	1.5710	9190	6175	0.0177	.00539

Table #13B - Circular Mils/Wire Gage Stranded Copper

Circular Mils/Wire Gage - Stranded Aluminum										
Conductor Size		No.of Wires	Wire Diameter Individual Strands		Class "B" Standard		Compressed Round		Compact Round	
AWG	Circ. Mils		mm	in	mm	in	mm	in	mm	in
12	6530	7	0.77	.0305	2.34	.092	2.26	.089		
10	10380	7	0.98	.0385	2.95	.116	2.87	.113		
8	16510	7	1.23	.0486	3.71	.146	3.61	.142	3.40	.134
6	26240	7	1.55	.0612	4.67	.184	4.55	.179	4.29	.169
4	41740	7	1.96	.0772	5.89	.232	5.72	.225	5.41	.213
3	52620	7	2.20	.0867	6.60	.260	6.40	.252	6.05	.238
2	66360	7	2.47	.0974	7.42	.292	7.19	.283	6.81	.268
1	83690	19(18)*	1.69	.0664	8.43	.332	8.18	.322	7.59	.299
1/0	105600	19(18)*	1.89	.0745	9.47	.373	9.19	.362	8.53	.336
2/0	133100	19(18)*	2.13	.0837	10.64	.419	10.31	.406	9.55	.376
3/0	167800	19(18)*	2.39	.0940	11.94	.470	11.58	.456	10.70	.423
4/0	211600	19(18)*	2.68	.1055	13.41	.528	13.00	.512	12.10	.475
250	MCM	37(35)*	2.09	.0822	14.61	.575	14.17	.558	13.20	.520
300	/kcmil	37(35)*	2.29	.0900	16.00	.630	15.52	.611	14.50	.570
350		37(35)*	2.47	.0973	17.30	.681	16.79	.661	15.60	.616
400		37(35)*	2.64	.1040	18.49	.728	17.93	.706	16.70	.659
500		37(35)*	2.95	.1162	20.65	.813	20.04	.789	18.70	.736
600		61(58)*	2.52	.0992	22.68	.893	22.00	.866	20.70	.813
750		61(58)*	2.82	.1109	25.35	.998	24.59	.968	23.10	.908
1000		61(58)*	3.25	.1280	29.26	1.152	28.37	1.117	26.90	1.060
1250		91	2.98	.1172	32.74	1.289	31.75	1.250		
1500		91	3.26	.1284	35.86	1.412	34.80	1.370		
1750		127	2.98	.1174	38.76	1.526	37.59	1.480		
2000		127	3.19	.1255	41.45	1.632	40.21	1.583		

NOTE:* Reduced number of wires for compact stranding are shown in parentheses

Table #14A - Circular Mils/Wire Gage Stranded Aluminum

Circular Mils/Wire Gage - Stranded Aluminum						
Conductor Size	Area		Approx. net Cable Weight		Nominal DC Resistance (25°C)	
AWG	mm²	sq.in.	kg per 1000 m	lbs per 1000 ft	Ohms/ 1000 m	Ohms/ 1000 ft
12	3.31	.00513	9.1	6.13	8.89	2.71
10	5.26	.00816	14.5	9.75	5.59	1.70
8	8.37	.01297	23.1	15.5	3.52	1.07
6	13.30	.02061	36.7	24.6	2.21	.674
4	21.15	.03278	58.3	39.2	1.39	.424
3	26.66	.04133	73.5	49.4	1.10	.336
2	33.63	.05212	92.7	62.3	0.875	.267
1	42.41	.06573	117	78.6	0.694	.211
1/0	53.51	.08291	147	99.1	0.550	.168
2/0	67.44	.1045	186	125	0.436	.133
3/0	85.03	.1318	234	157	0.346	.105
4/0	107.22	.1662	296	199	0.274	.0836
250	128.68	.1963	350	235	0.232	.0708
300	152.01	.2356	419	281	0.194	.0590
350	177.35	.2749	490	329	0.166	.0506
400	202.68	.3142	560	376	0.145	.0442
500	253.35	.3927	698	469	0.116	.0354
600	304.02	.4712	839	564	0.0967	.0295
750	380.03	.5890	1048	704	0.0774	.0236
1000	506.71	.7854	1396	938	0.0580	.0177
1250	633.38	.9817	1750	1174	0.0464	.0142
1500	760.06	1.1780	2100	1410	0.0387	.0118
1750	866.74	1.3740	2440	1640	0.0332	.0101
2000	1013.42	1.5710	2790	1880	0.0290	.00885

Table #14B - Circular Mils/Wire Gage Stranded Aluminum

Current Carrying Ratings of Cables and Cords

Both the NEC and CEC contain sections that state the maximum allowable current that a conductor may carry. These codes address the many variances, applications and factors that may be encountered including:

- Number of conductors within a cable or raceway
- Distances between conductors
- Ambient temperature in which the conductors will be placed
- Type of load that the conductor is connected to
- Rating of the overcurrent devices
- Conductor material type
- Type of insulation used on the conductor

All of these are crucial to the current rating of a conductor and represent only some of the many factors that must be taken into consideration when determining a conductor's current carrying capability.

Tables #15 and #16 list the allowable ampacities of copper and aluminum conductors when no more than three conductors are within a raceway or cable and the ambient temperature is 30°C or less. If there are more than three conductors or the ambient temperature is greater than 30°C, then correction factors must be used.

Insulation Type

Conductor insulation temperature ratings are designated by the manufacturer. Common temperature insulation values are 60°C, 75°C and 90°C. Ampacities of insulated conductors are based on an ambient temperature of 30°C. If the conductors are placed in environments that are above or below 30°C, then correction factors must be applied. These correction factors decrease the rating of a conductor if the temperature is above 30°C or increase the rating of a conductor if the ambient temperature is below 30°C.

The correction factors listed in table #17 are to be applied against the values of tables #15 and #16 if the temperature exceeds 30°C.

Size AWG kcmil	60 °C Type TW	75°C Types RW75, TW75	85 - 90°C Types R90, RW90,T90 Nylon, THHN, Paper, Mineral Insulated Cable
	Allowable Ampacities for Not More Than 3 Copper Conductors in a Raceway or Cable (Canada) Based on an Ambient Temperature of 30°C		
14	15	15	15
12	20	20	20
10	30	30	30
8	40	45	45
6	55	65	65
4	70	85	85
3	80	100	105
2	100	115	120
1	110	130	140
0	125	150	155
00	145	175	185
000	165	200	210
0000	195	230	235
250	215	255	265
300	240	285	295
350	260	310	325
400	280	335	345
500	320	380	395
600	355	420	455
700	385	460	490
750	400	475	500
800	410	490	515
900	435	520	555
1000	455	545	585
1250	495	590	645
1500	520	625	700
1750	545	650	735
2000	560	665	775

Table #15A - Ampacities for Copper Conductors (Canada)

Size AWG kcmil	60°C (140°F)	75°C (167°F)	85°C (185°F)	90°C (194°F)
	Types TW, UF	Types FEPW, RH, RHW, THHW, THW, THWN, XHHW, USE, ZW	Type V	Types TA, TBS, SA, SIS, FEP, FEPB, RHH, THHN, THHW, XHHW
18				14
16			18	18
14	20	20	25	25
12	25	25	30	30
10	30	35	40	40
8	40	50	55	55
6	55	65	70	75
4	70	85	95	95
3	85	100	110	110
2	95	115	125	130
1	110	130	145	150
1/0	125	150	165	170
2/0	145	175	190	195
3/0	165	200	215	225
4/0	195	230	250	260
250	215	255	275	290
300	240	285	310	320
350	260	310	340	350
400	280	335	365	380
500	320	380	415	430
600	355	420	460	475
700	385	460	500	520
750	400	475	515	535
800	410	490	535	555
900	435	520	565	585
1000	455	545	590	615
1250	495	590	640	665
1500	520	625	680	705
1750	545	650	705	735
2000	560	665	725	750

Allowable Ampacities for Not More than 3 Copper Conductors In Raceway or Cable (USA) Based on an Ambient Temperature of 30°C

Table #15B - Ampacities for Copper Conductors (USA)

Allowable Ampacities for Not More Than 3 Aluminum Conductors in a Raceway or Cable (Canada) Based on an Ambient Temperature of 30°C			
Size AWG kcmil	60 °C Type TW	75°C Types RW75, TW75	85 - 90°C Types R90, RW90, T90 Nylon, THHN, Paper, Mineral Insulated Cable
12	15	15	15
10	25	25	25
8	30	30	30
6	40	50	55
4	55	65	65
3	65	75	75
2	75	90	95
1	85	100	105
0	100	120	120
00	115	135	145
000	130	155	155
0000	155	180	185
250	170	205	215
300	190	230	240
350	210	250	260
400	225	270	290
500	260	310	330
600	285	340	370
700	310	375	395
750	320	385	405
800	330	395	415
900	355	425	455
1000	375	445	480
1250	405	485	530
1500	435	520	580
1750	455	545	615
2000	470	560	650

Table #16A - Ampacities for Aluminum Conductors (Canada)

Allowable Ampacities for Not More Than 3 Aluminum Conductors in a Raceway or Cable (USA) Based on an Ambient Temperature of 30°C				
Size AWG kcmil	60°C (140°F)	75°C (167°F)	85°C (185°F)	90°C (194°F)
	Types TW, UF	Types RH, RHW, THHW, THW, THWN, XHHW, USE	Type V	Types TA, TBS, SA, SIS, RHH, THHN, THHW, XHHW
12	20	20	25	25
10	25	30	30	35
8	30	40	40	45
6	40	50	55	60
4	55	65	75	75
3	65	75	85	85
2	75	90	100	100
1	85	100	110	115
1/0	100	120	130	135
2/0	115	135	145	150
3/0	130	155	170	175
4/0	150	180	195	205
250	170	205	220	230
300	190	230	250	255
350	210	250	270	280
400	225	270	295	305
500	260	310	335	350
600	285	340	370	385
700	310	375	405	420
750	320	385	420	435
800	330	395	430	450
900	355	425	465	480
1000	375	445	485	500
1250	405	485	525	545
1500	435	520	565	585
1750	455	545	595	615
2000	470	560	610	630

Table #16B - Ampacities for Aluminum Conductors (USA)

Correction Factors

Conductor Correction Factors				
Ambient Temperature		Correction Factor		
°C	°F	*60°C	*75°C	*85-90°C
40°C	104°F	0.82	0.88	0.90
45°C	113°F	0.71	0.82	0.85
50°C	122°F	0.58	0.75	0.80
55°C	131°F	0.41	0.65	0.74
60°C	140°F	—	0.58	0.67
70°C	158°F	—	0.35	0.52
75°C	167°F	—	—	0.43
80°C	176°F	—	—	0.30

*Note: Refers to Conductor Insulation Temperature Rating

Table #17 - Temperature Correction Factors

Ampacity Correction Factors for Tables #15 and #16	
Number of Current Carrying Conductors	Ampacity Correction Factor
1-3	1.00
4-6	0.80
7 - 24	0.70
25 - 42	0.60
43 and up	0.50

Table #18 - Ampacity Correction Factors

Conductor Material

Copper and aluminum are the two common materials used as electrical conductors. Copper is approximately 63% more efficient as a conductor and, therefore, may carry more current than aluminum.

Note: An aluminum conductor must be increased by two gage sizes to be comparable to a copper conductor.

A #10 AWG copper conductor has approximately the same current rating as a #8 AWG aluminum conductor. Refer to tables #15 and #16 and compare how closely the above gage rule works.

Adjacent Conductors

Each current carrying conductor is a source of additional heat. Placing more than three current carrying conductors in a raceway or cable will result in a derating of the conductors. Table #18 lists the derating factors for the conductors or cables of tables #15 and #16. The derating factors of table #18 should only be applied to current carrying power conductors.

Adjacent Conductors (cont'd)

Exceptions to the derating factors of table #18 are as follows:

1. Control wiring conductors.
2. Grounding or bonding conductors.
3. Neutral conductors except those neutral conductors that are connected to harmonic producing loads such as fluorescent lighting, high intensity discharge lighting, data processing equipment, VFD's and other non linear impedance loads which produce harmonics and cause harmonic currents to flow in the neutral of a 3ϕ 4W wye system.

A neutral conductor is defined as that conductor which carries the unbalanced load. Because a conductor is white and is grounded does not mean that it is a neutral. See Section One, pages 28 and 29, for additional information.

An example will help to explain the meaning of what is classified as a neutral.

Examples:

- A 120/208V 1ϕ 3W load has two ungrounded (line) conductors and one white wire. The white wire will be considered as a current carrying conductor and will not be classified as a neutral.

- A 120/240V 1ϕ 3W load has two ungrounded (line) conductors and one white wire. The white wire will be considered to be a neutral.

- A 120V 2W load has one ungrounded (line) conductor and one white wire. The white wire is considered to be a current carrying conductor and is not classified as a neutral.

Note: When determining conduit fill, all conductors must be counted. No conductors are exempt from being counted when determining the size of a conduit.

Other Factors (Direct Burial, Sheath Currents, Load Types)

Conductors that are placed underground require additional factors to determine their current carrying capabilities. Cable and wire manufacturers provide engineered data that should be consulted when specific directions are not available by local electrical authorities and codes.

Sheath-Type Conductor Cables

Single-conductor cables which have continuous sheaths of steel, lead, aluminum, or copper, and are carrying alternating current, will have a voltage induced in these metal sheaths, as shown in illustration #63.

If the the sheaths are connected so as to complete an electrical circuit, a circulating current called sheath current will flow, as shown in illustration #64.

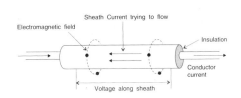

Illustration #63 - Sheath Voltage

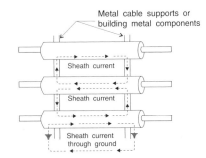

Illustration #64 - Sheath Current

Sheath-Type Conductor Cables (cont'd)

These sheath currents will raise the temperature of the cable and may cause the insulation of the individual conductors to be subjected to temperatures above their ratings. The need to derate single-conductor cables depends upon a number of factors such as spacing between each phase conductor, load current, isolation of the metal sheath at one end and the type of raceway. It is recommended that the cables be either:

1. Derated to 70% of their current carrying rating.
2. Derated in accordance with the manufacturers recommendations.
3. Installed in such a manner as to prevent the flow of sheath currents. To prevent sheath currents, it is necessary to ensure that all paths by which they may circulate are kept open. Cable sheaths should be grounded at the supply end only and thereafter be isolated from ground and from other cables.

Illustration #65 shows that isolation may be attained by mounting the cables on insulated supports thus effectively preventing the flow of sheath currents.

Isolation from ground and other cables can also be attained by installing the cables in individual ducts of insulating material or by employing cables jacketed with PVC or a similar type of insulation.

Note: The sheaths should be isolated from any metal enclosures or other terminations at the load side (see illustration #65). Also, a bonding conductor must now accomodate these single conductor cables because the sheaths can no longer be connected to ground or bonded to a metal enclosure.

A *final factor* that should be considered when determining a conductor's current rating is the load itself. Loads which are classified as continuous may require that published values of conductor ampacities be derated. Local code rules should be consulted and followed.

Sheath Type Conductor Cables (cont'd)

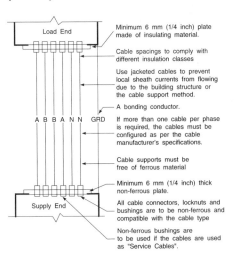

Minimum 6 mm (1/4 inch) plate made of insulating material.

Cable spacings to comply with different insulation classes

Use jacketed cables to prevent local sheath currents from flowing due to the building structure or the cable support method.

A bonding conductor.

If more than one cable per phase is required, the cables must be configured as per the cable manufacturer's specifications.

Cable supports must be free of ferrous material

Minimum 6 mm (1/4 inch) thick non-ferrous plate.

All cable connectors, locknuts and bushings are to be non-ferrous and compatible with the cable type

Non-ferrous bushings are to be used if the cables are used as "Service Cables".

Illustration #65 - Isolation of Cable Sheaths

Current Ratings Of Cords

Flexible cords have their own tables for conductor ampacities. Table #19A (Canada), and #19B (USA) contain the current ratings of various cords based on two or three current carrying copper conductors. If more than two or three current carrying conductors are in the cord, derating factors from table #18 will apply. Exceptions to the derating factors of table #18 are the same as those stated under the heading of adjacent conductors (see page 95).

Current Ratings Of Solid Busbars

Solid copper or aluminum busbars are used in panelboards, MCCs, switchgear and other electrical distribution equipment. Tables #20 and #21 list the current carrying ratings of commonly used busbar sizes. The values shown are conditional and actual ratings may vary depending on busbar location, configuration and application.

Allowable Ampacity of Flexible Cords (Canada) Based on an Ambient Temperature of 30°C (86°F)						
Size AWG	Tinsel Cords Types TPT TST	Christmas Tree Cord Types CXWT PXT	Types E, EO, ETT	Types PXWT, SV, SVO, SJ, SJO, SJOW, S, SO, SOW, SPT-1, SPT-2, SPT-3, SVT, SJT, SJTW, ST, STW		Types HSJO, HPN, DRT
				2 Current-Carrying Conductors	3 Current-Carrying Conductors	
27	0.5	—	—	—	—	—
26	—	—	—	—	—	—
24	—	—	—	—	—	—
22	—	—	—	—	—	—
20	—	2	—	2	—	—
18	—	5	5	10	7	10
16	—	7	7	13	10	15
14	—	—	15	18	15	20
12	—	—	20	25	20	25
10	—	—	25	30	25	30
8	—	—	35	40	35	40
6	—	—	45	55	45	50
4	—	—	60	70	60	60
2	—	—	80	95	80	—

Table #19A - Allowable Ampacity for Cords (Canada)

Allowable Ampacity of Flexible Cords (USA)
Based on an Ambient Temperature of 30°C (86°F)

Size AWG	Thermoset Type TS / Thermo plastic Types TPT, TST	Thermoset Types C, E, EO, PD, S, SJ, SJO, SJOO, SO, SOO, SP-1, SP-2, SP-3, SRD, SV, SVO, SVOO / Thermoplastic Types ET, ETT, ETLB, SE, SEO, SJE, SJEO, SJT, SJTO, SJTOO, SPE-1, SPE-2, SPE-3, SPT-1, SPT-2, SPT-3, ST, STO, STOO, SRDE, SRDT, SVE, SVEO, SVT, SVTO, SVTOO		Types AFS, AFSJ, HPD, HPN, HS, HSJ, HSJO, HSO	Asbestos Types AFC AFPD, AFPO
		A*	B*		
27	0.5	—	—	—	—
20	—	5	7	—	—
18	—	7	10	10	6
17	—	—	12	—	—
16	—	10	13	15	8
15	—	—	—	17	—
14	—	15	18	20	17
12	—	20	25	30	23
10	—	25	30	35	28
8	—	35	40	—	—
6	—	45	55	—	—
4	—	60	70	—	—
2	—	80	95	—	—

NOTE: A* - 3 Current-Carrying conductors, B* - 2 Current-Carrying Conductors in each Cord

Table #19B - Allowable Ampacity for Cords (USA)

Solid Copper Busbars (Based on a 25°C Ambient Temperature)				
Size	**Weight**	**Area**	***Current Carrying Rating/A**	
inches	lb/ft	cmil	****Horizontal**	Vertical
¼" × 1"	0.96	318,309	530	432
¼" × 1 ½"	1.44	477,463	740	605
¼" × 2"	1.93	636,618	940	765
¼" × 2 ½"	2.41	795,772	1150	935
¼" × 3"	2.89	954,927	1335	1060
¼" × 3 ½"	3.38	1,114,082	1500	1220
¼" × 4"	3.86	1,273,236	1700	1385
¼" × 5"	4.82	1,591,545	2025	1630
¼" × 6"	5.78	1,909,854	2350	1930
¼" × 7"	6.76	2,228,163	2675	2175

Notes:
* Values are based on 1 busbar per phase, 600 V, 25°C. If two or three busbars per phase are used, a derating factor of approx. 0.75 must be applied.
** Busbars that are oriented horizontally dissipate heat more effectively

Table #20 - Solid Copper Busbars

Solid Aluminum Busbars (Based on a 25°C Ambient Temperature)				
Size	**Weight**	**Area**	***Current Carrying Rating/A**	
inches	lb/ft	cmil	****Horizontal**	**Vertical**
¼" × 1"	0.292	318,309	412	335
¼" × 1 ½"	0.438	477,463	567	462
¼" × 2"	0.587	636,618	722	588
¼" × 2 ½"	0.733	795,772	878	715
¼" × 3"	0.879	954,927	1012	825
¼" × 3 ½"	1.03	1,114,082	1150	935
¼" × 4"	1.17	1,273,236	1297	1055
¼" × 5"	1.47	1,591,545	1580	1285
¼" × 6"	1.76	1,909,854	1835	1495
¼" × 7"	2.10	2,228,163	2100	1700

Notes:
* Values are based on 1 busbar per phase, 600 V, 25°C. If two or three busbars per phase are used, a derating factor of approx. 0.75 must be applied.
** Busbars that are oriented horizontally dissipate heat more effectively

Table #21 - Solid Aluminum Busbars

Voltage Drop

In addition to the many factors considered under the heading of current ratings, the size of a current carrying conductor must also stay within the allowable or recommended voltage drop limits. No conductor can deliver load current without experiencing a voltage drop (line drop IR) and expending heat (line loss I^2R). Both line drop and line loss are unwelcome byproducts of transporting electrical current to a load. Solutions to reduce these two values vary and depend upon the practicality, economics and extent of the problem.

Examining the two formulas, I^2R and IR, reveals that two approaches may be taken to minimize the losses. First the load current may be reduced and the line voltage increased. This is typically done by electrical utility companies which generate at a lower voltage, then step-up the voltage through a transformer and transport the load current over great distances at high voltages. The voltage is reduced at the point of utilization to match the load.

This method is both practical and economical for the utility company and is commonly practiced.

The second method involves reducing the impedance value of the conductor. Local codes and recommended wiring practices state that no more than a 5% voltage drop may occur from the point of electrical service to the point of utilization (load). It is also recommended that no more than a 3% voltage drop occur from the point of distribution (a circuit breaker for example) to the load.

Voltage Drop Factors

It would be very convenient and relatively simple if the only factor to consider in determining voltage drop was the resistance of the wire. DC electrical circuits contain only DC resistance and involve the least complications when calculating voltage drop.

Voltage Drop Factors (cont'd)

AC electrical circuits contain three types of opposition to the flow of current. They are known as AC resistance (Rac), inductive reactance (X_L) and capacitive reactance (X_C).

Rac is a function of Rdc, eddy currents, hysteresis, skin effect and dielectric absorption. At 60 Hz and conductor sizes of 300 kcmil or smaller, Rac is commonly considered to be equal to Rdc. Skin effect becomes more significant at 60 Hz with conductors larger than 300 kcmil. Manufacturers' literature should be consulted for resistance values.

The effect of inductive reactance (X_L) can be neglected unless the size of wire exceeds those given in table 22. If the circuits inductive reactance is a consideration, the simplest solution is to bring the current carrying conductors closer together. The effect of capacitive reactance (X_C) can be neglected except for long distance transmission lines.

Calculating Voltage Drop

Assuming that X_L and X_C are negligible and that Rac is approximately Rdc, the formulas to calculate total voltage drop(VD) are shown in table #23.

Conductor Spacings and Inductive Reactance											
The effect of inductance can be neglected unless the size of wire exceeds the following:											
	Size of wire with the following spacings										
Type of Load	In conduit	2 1/2 in	4 in	5 in	6 in	8 in	12 in	18 in	24 in	36 in	48 in
Incandescent Lamps: 60 Hz	4/0	2/0	1/0	1/0	1	1	2	2	2	3	3
Incandescent Lamps: 25 Hz	600 kcmil	400 kcmil	300 kcmil	4/0	4/0	4/0	3/0	3/0	3/0	3/0	2/0
Motors: 60 Hz	1	3	4	4	5	5	5				

Table #22 - Conductor Spacings & Inductive Reactance

Calculating Voltage Drop	
Type of Circuit	**Formula**
1. Two wire dc	$VD = 2\rho \times I \times L/A$
2. Three wire dc	same as 1
3. Two wire ac	same as 1
4. Three wire ac single phase	same as 1
5. Three wire ac two phase	$VD = 2\rho \times I \times L \times 0.85/A$
6. Three wire ac three phase	$VD = 2\rho \times I \times L \times 0.866/A$
7. Four wire ac three phase	$VD = 2\rho \times I \times L \times 0.866/A$

Notes:

a. ρ is the resistivity of the conductor. ρ for copper is 10.4 Ω/cmil - foot at 20°C and ρ for aluminum is 17.1Ω/cmil - foot at 20°C.

b. I is the largest load current present in the circuit being considered and is expressed in amperes.

c. L is the one-way length of the conductor from the source to the load expressed in feet.

d. A is the cross sectional area of the conductor expressed in cmil. Tables 11 to 14 list areas of conductors expressed in cmil.

e. 2 is a multiplier used to account for the fact that the circuit involves both a supply conductor and a return conductor. (It is assumed that the return conductor is NOT grounded at either the load or the source).

f. 0.85 and 0.866 are multipliers that account for the unique configuration of the circuit and are only used as noted in 5, 6 and 7.

Table #23 - Voltage Drop Formulas

Calculating Voltage Drop (cont'd)
Example 1:

Determine the voltage drop of a branch circuit supplying 23 A to a two wire AC 120 V load. The branch conductors are copper size #10 AWG and the load is 50 feet from the panelboard. Does this circuit meet the recommended maximum 3% VD?

Solution:

From table #23, the formula is:

$VD = 2\rho \times I \times L/A$

Substituting for the actual circuit values results in:

$VD = 2 \times 10.4 \times 23 \times {}^{50}/10380$
$VD = 2.3$ V

Maximum allowable or recommended voltage drop is 3% of 120 V = .03 x 120 = 3.6 V.

Therefore, this circuit meets design requirements.

Example 2:

Determine the voltage drop of a branch circuit supplying 23 A to a three wire three phase 480 V load. The branch conductors are copper size #10 AWG and the load is 400 feet from the panelboard. Does this circuit meet the recommended maximum 3% VD and if not what are the options?

Solution:

$VD = 2 \times 10.4 \times 23 \times 400 \times {}^{0.866}/10380$
$VD = 15.97$ V

Maximum allowable or recommended voltage drop is 3% of 480V = .03 x 480 = 14.4 V.

This circuit is not in compliance with code and a larger conductor than #10 AWG is required.

Note: If inductive reactance is a potential factor, consult engineering handbooks for multiplication factors.

Conductor Markings, ID, Colors

All approved cables and conductors must have standard markings on them. Typical markings consist of the following:

- maximum voltage rating
- type of insulation
- manufacturers name, logo or trademark
- AWG size or circular-mil area

Conductors are identified by color and color groupings according to their function. Table #24 lists the colors and the conductor function.

HV Conductor Design

Conductors and cables that are designed to operate at 2000 V or above may be shielded as shown in illustration #66.

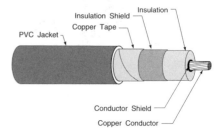

Illustration #66 - Conductor Example (2000 V & Over)

Color Identification for Conductors	
Function	**Color**
Grounded conductor	white or natural grey
Grounding conductor	bare, green, green with yellow stripes
Ungrounded conductor	black, blue, red, or any other color except as mentioned above.

Table #24 - Conductor Color Identification

HV Conductor Design (cont'd)

The NEC and CEC codes require that all cables at 5 kV and above are to be shielded. Cables rated at 2 - 5 kV may also need to be shielded depending on the following factors:

1. need to ensure safety of personnel
2. use of single conductors in wet locations
3. direct earth burial of conductors
4. exposure level of conductors to conductive materials (metal dusts etc.)

Purpose of Shielding

The shielding of a conductor or cable confines the electric field to the insulation surrounding the conductor. This is accomplished by using both conducting and semiconducting materials that surround the inner and outer surfaces of the insulation. This serves to limit the electric field to the space between the shield and the conductor. The insulation shield has the following purposes:

1. To confine the electric field within the cable.
2. To equalize voltage stress within the insulation.
3. To shield the cable from induced potentials and minimize the flow of circulating currents.
4. To limit both electromagnetic and electrostatic interference.
5. To reduce electric shock.

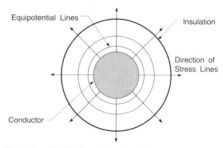

Illustration #67 - HV Conductor Stress Lines

HV Conductor Design (cont'd)

Illustration #67 shows the voltage stresses that exist in a HV conductor. Each of the rings shown represents an equipotential line, that is, a line of equal electrical pressure measured in volts. The further the distance from the conductor to the insulation the lower the value of electrical pressure (volts/mm). If the conductor is perfectly round with no imperfections such as scratches, nicks or thicknesses, the equipotential lines will be uniform.

If stranded cable is used, as shown in illustration #68, the uniformity of the equipotential lines will vary. To restore uniformity, a shielding tape or semiconducting tape is used to fill in the gaps. It is necessary to have this uniformity to ensure that the insulation covering the conductors will not have different values of stress. Voltage stress variances may lead to insulation failure. Voltage stress variances may also be present due to air voids in the cable insulation.

Illustration #69 indicates how a semiconducting tape has restored uniformity to the conducting surface and has thereby eliminated variances in stress points on the insulation.

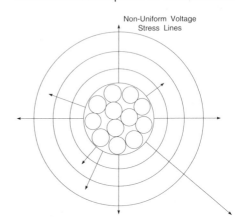

Non-Uniform Voltage Stress Lines

Illustration #68 - Non-uniform Stress Lines of Stranded Conductors

HV Conductor Design (cont'd)

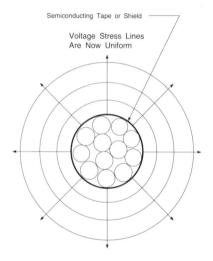

Semiconducting Tape or Shield

Voltage Stress Lines Are Now Uniform

Illustration #69 - Semiconducting Tape & Stress Uniformity

Illustration #70 includes the outer shield which may be a continuous series of copper braids, copper tape or copper corrugated tube. The purpose of this shield is to establish uniformity of electrical stress in reference to external elements such as the ground plane or another conductor.

Shielding to Avoid Corona

As operating voltages increase normal insulation may prove to be inadequate. Electrical discharges called CORONA will deteriorate insulation and cause it to break down and fail. Corona is defined as the breakdown of air particles caused by a voltage or potential difference which is greater than the air can withstand. Corona produces ozone, heat and charged particles called ions which will quickly deteriorate the insulation. Corona will form in the air voids of insulated cables, splices or terminations. Shielding will help to eliminate and/or reduce corona.

Shielding to Avoid Corona (cont'd)

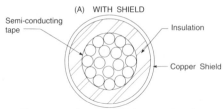

(A) WITH SHIELD

Semi-conducting tape

Insulation

Copper Shield

The purpose of copper shield is to equalize all the equipotential lines as referenced to earth or to another potential.

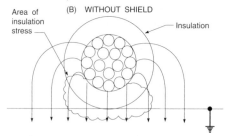

(B) WITHOUT SHIELD

Area of insulation stress

Insulation

Illustration #70 - Effect of Outer Tape Shield for Stress Uniformity

Terminating and Splicing

There are three common methods to terminate or splice shielded cables:

1. no stress control

2. tape-constructed stress cone

3. heat-shrinkable stress cone.

Methods 2 and 3 are necessary to ensure a safe, reliable and approved method of terminating and splicing cable. Certain precautions and procedures should be followed in preparing the cable for splicing:

Terminating and Splicing (cont'd)

1. Careful removal of all cable layers.
2. Careful removal of insulation surrounding the conductor.
3. Cut back layers to prescribed distances.
4. Cleaning of all components, layers and parts.

Illustration #71 is an example of a HV splice. Illustration #72 is an example of a HV termination.

Manufacturers' literature should be consulted when using a splice or termination kit. Following the recommendations contained within the kit will assure a reliable and safe installation.

A number of factors can lead to a splice or termination failure. Table #25 lists failures and probable causes.

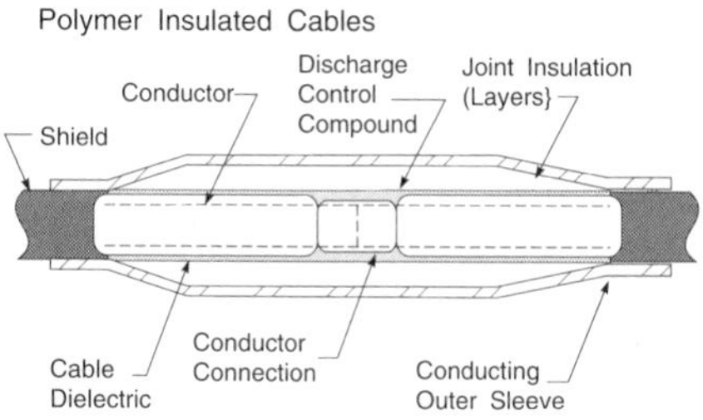

Polymer Insulated Cables

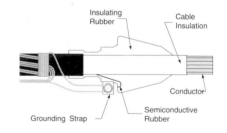

Illustration #71 - HV Splice

Illustration #72 - HV Termination (Pre-fabricated Stress Cone)

Termination Failures and Causes	
Problem	**Cause/Remedy**
Damaged cable ends	Protect cable ends when pulling through duct systems. Keep cable ends sealed to prevent entrance of moisture.
Damage to insulation	Bending the cable too sharply. Follow manufacturers' published cable bending values.
Weakened insulation	Accidental cuts into cable insulation and air voids during the removal of the jacket or outer shielding.
High stress concentration and insulation failure	Improper cut-off of the metallic or semiconductive shielding leaving strands that point inward, or leaving an uneven edge on the metallic shielding.
Interface or connection failure	Failure to remove or improper removal of semi-conductive material from the cable insulation.
Particles of metal imbedded in the insulation	Using conducting abrasives for cleaning the insulation. Use only non-conducting abrasives for cleaning.
Weakened insulation	Excessive use of solvents when cleaning cable insulation. Keep solvents away from semiconductive tapes or extruded layers in the cable.
Air voids	Poor pencilling of the cable insulation. Pencilling must be smooth and round.
Damaged conductor strands and loss of current carrying ability and the build up of heat	Nicking the conductors when removing insulation.
Improper or loose connectors	Connectors should have smooth outer surfaces and be crimped or soldered.

Table #25 - Termination and Splicing Failures

SECTION THREE

RACEWAYS

Raceway Types

A raceway may be defined as an enclosed channel that is designed to hold wires, cables or busbars. The following are all classified as raceways:

- rigid metallic conduit
- rigid nonmetallic conduit
- electrical metallic tubing
- flexible metallic and nonmetallic conduit
- nonmetallic tubing
- cable trays
- busways
- cablebus
- wireways
- underfloor raceways
- cellular floors
- surface raceways

Note: Before specifying, designing or installing raceway systems, refer to NEC articles 345-365 (USA) or to CEC section 12 (Canada).

Rigid Metallic Conduit

There are several significant benefits and advantages to rigid metallic conduit:

a. A high degree of mechanical protection from impact and crushing.
b. Resistive to chemical action of cement when embedded in concrete.
c. Flexibility (adding or withdrawing wire from the conduit).
d. Excellent bonding and grounding path.
e. Available in steel, aluminum, silicon bronze alloy and plastic coated.
f. Suitable for bending, cutting, coupling and threading.

There are two classes of rigid metallic conduit: magnetic (steel) and nonmagnetic (aluminum and silicon bronze alloy). The nonmagnetic conduits could also be classified as being corrosion resistant depending upon the environment in which they are placed.

Steel Conduit

Rigid steel conduit may be used when severe corrosive conditions are not present. Steel is much harder, stronger and heavier than its aluminum counterpart. There are three types of rigid steel conduit:

1. Galvanized: Most often used for concrete encasement, wet locations and when exposed to outdoor conditions.

2. Enameled: Most often used for interior wiring.

3. Sheradized: Coated with a green film that is somewhat corrosion resistant.

Rigid steel conduit also comes in a plastic coated form. This steel conduit has a PVC seamless jacket or coating approximately 40 mils thick.

Rigid Steel Conduit							
Nominal Trade Size		Outside Diameter inches	Inside Diameter inches	Wall Thickness	Threads per inch	Weight lbs.per 100 ft	Weight kg per m
inches	mm						
½	15	0.840	0.622	0.109	14	79	1.18
¾	20	1.050	0.824	0.113	14	105	1.56
1	25	1.315	1.049	0.133	11½	153	2.28
1¼	32	1.660	1.380	0.140	11½	201	3.00
1½	40	1.900	1.610	0.145	11½	249	3.71
2	50	2.375	2.067	0.154	11½	332	4.95
2½	61	2.875	2.469	0.203	8	527	7.85
3	75	3.500	3.068	0.216	8	683	10.18
3½	90	4.000	3.548	0.226	8	831	12.38
4	100	4.500	4.026	0.237	8	972	14.49
5	125	5.563	5.047	0.258	8	1313	19.57
6	150	6.625	6.065	0.280	8	1745	26.00

Table #26 - Rigid Steel Conduit

Steel Conduit (cont'd)

The PVC jacket is corrosion resistant, flame retardant, oxide free, impact and shrinkage resistant and impervious to abrasion. Applications for this jacketed conduit include fertilizer and chemical plants, meat-packing plants, canneries and some petroleum refineries. See table #26 for dimensions and weights of steel conduits.

Aluminum Conduit

Aluminum rigid conduit is useful in areas where there are chemical fumes or vapors present that affect steel but not aluminum. An example would be a bulk fertilizer storage depot where the ammonium nitrate is corrosive to steel but not to aluminum. Aluminum is commonly used due to its light weight and the ease with which it can be handled, thereby lowering labor cost.

Maintenance costs are minimized because no painting, cleaning or protective coatings are required to prevent oxidization from occuring. A natural thin oxide layer forms on the aluminum and prevents any further corrosion. Locations where aluminum may not be used include:

a. Direct earth burial where moisture, chemicals in the earth and stray currents lead to severe aluminum corrosion.

b. Encasement in concrete treated with chlorides.

Aluminum, being nonmagnetic, will result in lower AC reactance values being experienced by the copper or aluminum wires that are inserted into it. This lower reactance results in as much as a 20% reduction in voltage drop compared to a steel conduit installation. See table #27 for dimensions and weights of aluminum conduits.

Aluminum Conduit (cont'd)

Rigid Aluminum Conduit				
Nominal Trade Size		Outside Diameter inches	Inside Diameter inches	Weight lbs per 100 ft	Weight kg per m
inches	mm				
½	15	0.839	0.622	27.94	0.416
¾	20	1.051	0.822	37.84	0.564
1	25	1.315	1.047	55.88	0.832
1¼	32	1.661	1.382	75.90	1.13
1½	40	1.901	1.610	91.74	1.37
2	50	2.374	2.067	124.74	1.86
2½	61	2.874	2.468	194.7	2.90
3	75	3.5	3.067	264.0	3.93
3½	90	4.0	3.547	314.6	4.69
4	100	4.5	4.028	374.0	5.57

Table #27 - Rigid Aluminum Conduit

Silicon Bronze Alloy Conduit

This conduit has a special corrosive resistance quality and is primarily used for installations that are exposed to the weather.

This includes oil refineries, chemical plants, effluent plants, underwater wiring (for example, between deck boxes and wet niche luminaires) and bridges and piers along the seacoast.

Conduit Threading

Rigid metallic conduits must be threaded, and the thread must be tapered. The Electrical Code requires that the threads should not be longer than maximum specified amounts as stated in table #28.

Conduit Bends

Field bends must be made without damaging or distorting the raceway. Illustration #73 contrasts the difference in appearance between an acceptable and an unacceptable bend.

Metallic Conduit Threads				
Size inches	Threads per inch	*Thread length L_T inches	Effective Length L_E inches	E_P Pitch diameter inches
½	14	0.78	0.53	0.758
¾	14	0.79	0.55	0.968
1	11.5	0.98	0.68	1.214
1¼	11.5	1.01	0.71	1.557
1½	11.5	1.03	0.72	1.796
2	11.5	1.06	0.76	2.269
2½	8	1.57	1.14	2.720
3	8	1.63	1.20	3.341
3½	8	1.68	1.25	3.838
4	8	1.73	1.30	4.334
4½	8	1.78	1.35	4.831
5	8	1.84	1.41	5.391
6	8	1.95	1.51	6.446

Table #28 - Conduit Threading

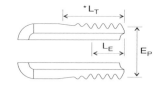

Conduit Bends (cont'd)

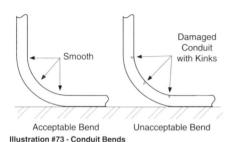

Illustration #73 - Conduit Bends

To ensure that conduit field bends are made without damage, a minimum bending radius must be maintained. Table #29 provides a listing of minimum bending radii.

Rigid metal conduits may be shaped, bent or formed by either using a manual bender, or a power assisted bender, as shown in illustration #74.

Minimum Radius for Field Bent Conduit (Inches)		
Size of Conduit inches	Conductors without lead sheath	Conductors with lead sheath
½	4	6
¾	5	8
1	6	11
1¼	8	14
1½	10	16
2	12	21
2½	15	25
3	18	31
3½	21	36
4	24	40
5	30	50
6	36	61

Table #29 - Minimum Radii for Conduit Bends

Conduit Bends (cont'd)

Conduits from ½ to 1 inch (13 to 25.4 mm) may be manually bent using a foot bender as shown in illustration #74A. Power assisted benders (illustrations #74B and #74C) are required for conduits larger than 1 inch (25.4 mm).

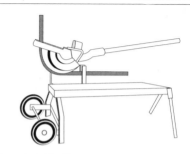

Illustration #74B - Mechanical Conduit Bender

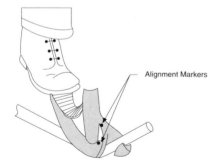

Alignment Markers

Illustration #74A - Manual Conduit Bender

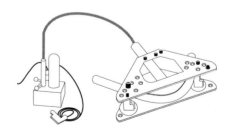

Illustration #74C - Hydraulic Conduit Bender

Conduit Bends (cont'd)

Illustration #75 displays typical on site conduit bends. Each of these bends requires that the electrician follow proper bending procedures as outlined by the bender manufacturer.

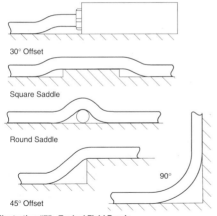

30° Offset

Square Saddle

Round Saddle

45° Offset

90°

Illustration #75 - Typical Field Bends

Cutting, Reaming and Threading

There are four stages involved in getting a piece of rigid metal conduit ready for installation:

1. shaping or bending
2. cutting
3. reaming
4. threading

In the second stage, conduit may be cut with either a hacksaw or a pipe cutter. Both methods are shown in illustration #76.

The third stage requires reaming the conduit and ensuring an abrasive-free inner tube surface. If the conduit has been cut using a hacksaw, the conduit will have a sharp burr that can damage the insulation of the conductors being drawn into it. If the conduit has been cut using a pipe cutter, a thick burr will have developed and extensive reaming will be necessary. A power vise as shown in illustration #77 will make conduit reaming much better, and also easier.

Cutting, Reaming, Threading (cont'd)

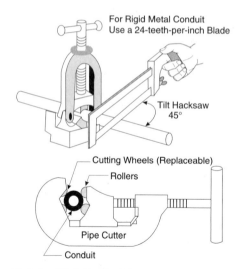

For Rigid Metal Conduit
Use a 24-teeth-per-inch Blade

Tilt Hacksaw
45°

Cutting Wheels (Replaceable)

Rollers

Pipe Cutter

Conduit

Illustration #76 - Cutting Conduit

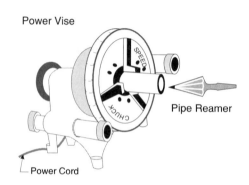

Power Vise

SPEED

CHUCK

Pipe Reamer

Power Cord

Illustration #77 - Power Reaming

Threading the conduit is the final stage of preparing a conduit. The size and length of thread must conform to NEC and CEC codes (see table #28). The conduit should be supported in a vise, as shown in illustration #78.

Cutting, Reaming, Threading (cont'd)

Modern conduit threaders produce a tapered, rather than a paralleled thread. To ease the effort and prolong the life of the threading die, a cutting liquid lubricant should be liberally applied whenever threading is done.

Illustration #79 displays the most common configurations for conduit bodies. These conduit bodies may contain splices, but certain code restrictions do apply (see NEC 370-6(c)).

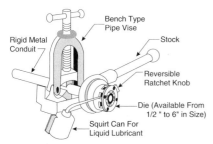

Illustration #78 - Threading Conduit

Conduit Bodies/Condulets

Threaded conduit bodies or condulets are commonly used to provide tap off or transition points in the wiring system.

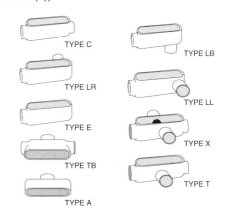

Illustration #79 - Common Condulets

Conduit Supports

Canada and the USA differ in their spacing requirements for supporting rigid metal conduit.

Refer to table #30A for Canadian support standard and table #30B for the USA standard.

Conduit Support Spacings (Canada)		
Conduit size inches	Maximum Distance Between Supports	
	feet	m
½	5	1.5
¾	5	1.5
1	6.5	2
1¼	6.5	2
1½	10	3
2	10	3
2½	10	3
3 and larger	10	3

Table #30A - Conduit Support Spacings (Canada)

Conduit Support Spacings (USA)		
Conduit size inches	Maximum Distance Between Supports	
	feet	m
½	10	3
¾	10	3
1	12	3.7
1¼	14	4.3
1½	14	4.3
2	16	4.9
2½	16	4.9
3 and larger	20	6.1

Table #30B - Conduit Support Spacings (USA)

Conduit Supports (cont'd)

A variety of support straps are available for rigid metallic conduits, some of which are shown in illustration #80.

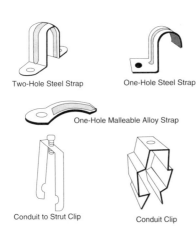

Two-Hole Steel Strap

One-Hole Steel Strap

One-Hole Malleable Alloy Strap

Conduit to Strut Clip

Conduit Clip

Illustration #80 - Conduit Supports

Rigid Nonmetallic Conduit

Rigid nonmetallic conduits are grouped into three categories:

1. **Thin Wall** known as type A in the USA and called EB1 in Canada.
2. **Heavy Wall** known as type 40 in the USA and called RE, and DB2/ES2 in Canada.
3. **Extra Heavy Wall** known as type 80 in the USA and called rigid PVC and HFT in Canada.

Materials that are used for making rigid nonmetallic conduits include: fiber, soapstone, fiberglass epoxy, high density polyethylene and polyvinyl chloride (PVC). Table #31A provides a brief description and application for each classification in Canada. Table #31B provides a brief description and application for each of the conduit classifications in the USA.

Rigid Nonmetallic Conduit (cont'd)

Conduit Classifications (Canada)		
Rigid nonmetallic conduit	Description	Application
PVC	Conduit made of un-plasticized polyvinyl chloride	Wet or Dry location both above and below ground or in concrete
HFT	Halogen free thermoplastic	Same as PVC
RE	Reinforced thermoset material	Direct burial or concrete encasement
DB2/ES2	Heavy Wall PVC	Direct burial or concrete encasement
EB1	Thin Wall PVC	Concrete encasement only

Table #31A - Rigid Nonmetallic Conduit (Canada)

Conduit Classifications (USA)		
Rigid nonmetallic conduit	Description	Application
Type A	Thin Wall	Concrete encasement only
Type 40	Heavy Wall	Direct burial or concrete encasement
Type 80	Extra Heavy Wall	Direct burial, concrete encasement and above-ground installation

Table #31B - Rigid Nonmetallic Conduit (USA)

Type 80 (PVC)

PVC has excellent physical, mechanical and electrical characteristics. It has both high tensile and high impact strength. PVC is half the weight of aluminum and one-sixth the weight of steel, therefore reducing shipping charges and installation costs.

PVC does not require threading, is easily cut and requires minimal reaming. Coupling lengths of PVC conduit together is easily accomplished with a solvent cement and no special fittings, such as unions or Erickson couplings, are required.

PVC Conduit Dimensions						
Nominal Size		Outside Diam. inches	Inside Diam. inches	Wall Thickness inches	Weight lbs per 100 ft	Weight kg per m
inches	mm					
½	15	0.840	0.622	0.109	16.0	.238
¾	20	1.050	0.824	0.113	21.3	.317
1	25	1.315	1.049	0.133	31.6	.471
1¼	32	1.660	1.380	0.140	42.9	.639
1½	40	1.900	1.610	0.145	52.1	.776
2	50	2.375	2.067	0.154	71.0	1.06
2½	61	2.875	2.469	0.203	112.5	1.68
3	75	3.500	3.068	0.216	147.2	2.19
3½	90	4.000	3.548	0.226	177.0	2.64
4	100	4.500	4.026	0.237	210.0	3.13
5	125	5.563	5.047	0.258	284.3	4.24
6	150	6.625	6.065	0.280	369.4	5.51

Table #32 - PVC Conduit Data

Type 80 (PVC) (cont'd)

Drawing wire into the PVC raceway requires much less effort due to its smooth interior finish. PVC is resistant to corrosion and, therefore, requires no maintenance or painting. PVC is non-sparking, non-magnetic and non-conductive therefore eliminating shock hazard, voltage drop impediments and heat losses.

PVC is resistive to acids, alkalies, salt solutions, chlorine and many other chemicals. PVC also has the ability to withstand exposure to direct sunlight, heat (a maximum of 75°C/167°F) and weather. PVC will not support combustion. Table #32 contains data on sizes and weights of PVC conduit.

Handling PVC

Illustration #81 shows the ease of working with PVC. PVC may be cut with a hacksaw and then reamed with a pen knife to remove the plastic burrs.

Coupling two conduits together or connecting a length of conduit to a box or fitting is easily accomplished by the use of a solvent cement. **Solvent welds appear to harden or set up instantly but actually require 24 hours to cure properly**.

PVC is a thermoplastic material and lends itself to being reshaped with the application of heat. A flameless heat source is recommended, as shown in illustration #81. The recommended heat source is either a hot air stream from an electric hand unit or an infrared propane unit. The necessary temperature for bending PVC pipe properly is 260°F (126.6°C). This heat must be applied evenly over an area approximately 10 times the diameter of the pipe before any attempt is made to bend it. Over or under heating the pipe will result in deformations or kinks in the conduit.

Type 80 (PVC) (cont'd)

PVC conduits require adequate support. Maximum spacings between conduit supports are listed in table #33.

Note: Supporting straps should not be tightened firmly in order to allow for lineal movement (expansion and contraction) caused by temperature changes.

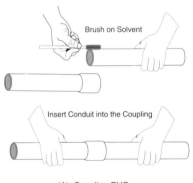

Brush on Solvent

Insert Conduit into the Coupling

(A) Coupling PVC

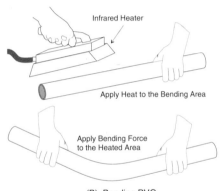

Infrared Heater

Apply Heat to the Bending Area

Apply Bending Force to the Heated Area

(B) Bending PVC

Illustration #81 - Joining and Bending PVC Conduit

Type 80 (PVC) (cont'd)

PVC Conduit Support Spacings			
Nominal Size		Maximum Spacing Between Supports	
inches	mm	feet	m
½	15	2.5	0.76
¾	20	2.5	0.76
1	25	2.5	0.76
1¼	32	4	1.22
1½	40	4	1.22
2	50	5	1.52
2½	61	6	1.83
3	75	6	1.83
3½	90	7	2.13
4	100	7	2.13
5	125	7	2.13
6	150	8	2.44

Table #33 - PVC Conduit Support Spacings

Electrical Metallic Tubing (EMT)

EMT is about half as thick as rigid metallic conduit and, therefore, cannot be threaded. Due to the ease with which it can be handled and installed, EMT is widely used and is a popular means of wiring. EMT makes use of threadless couplings and connectors, as shown in illustration #82, thus allowing the installer to work unimpeded in awkward locations. EMT is light in weight and is relatively easy to bend by hand. EMT is available in galvanized steel (see table #34), aluminum and silicon bronze alloy.

Bending requirements and conduit body types are the same for EMT as for rigid metallic conduit (see table #29 and illustration #79).

Electrical Metallic Tubing (EMT)
(cont'd)

Galvanized Steel Conduit (EMT)					
Size	Internal Diameter	External Diameter	Wall Thickness	Weight per 100 ft	Weight per m
inches	inches	inches	inches	pounds	kg
½	0.622	0.706	0.042	28.5	.425
¾	0.824	0.922	0.049	43.5	.648
1	1.049	1.163	0.057	64.0	.951
1¼	1.380	1.510	0.065	95.0	1.416
1½	1.610	1.740	0.065	110.0	1.639
2	2.067	2.197	0.065	140.0	2.086
2½	2.731	2.875	0.072	205.0	3.055
3	3.356	3.500	0.072	250.0	3.726
4	4.334	4.500	0.083	370.0	5.514

Table #34 - Galvanized EMT

EMT Conduit Support Spacings (Canada)		
Trade Size of EMT	Maximum Spacing Between Supports	
inches	feet	m
½ and ¾	5	1.5
1 and 1¼	6.5	2.0
1½ and Larger	10	3.0

Table #35 - EMT Support Spacings (Canada)

EMT must be supported at regular intervals by approved straps with maximum spacings, as listed in table #35. In the USA EMT may be supported at a maximum distance of 10 feet (3m); and in both USA and Canada, EMT must be supported within 3 feet (1m) of any outlet box, junction box, cabinet or fitting.

Electrical Metallic Tubing (EMT) (cont'd)

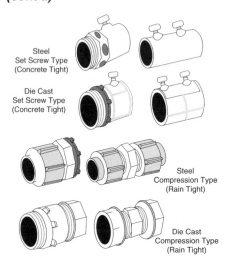

Steel
Set Screw Type
(Concrete Tight)

Die Cast
Set Screw Type
(Concrete Tight)

Steel
Compression Type
(Rain Tight)

Die Cast
Compression Type
(Rain Tight)

Illustration #82 - EMT Threadless Connectors and Couplings

Conduit Fill

The physical size of a conduit and the mutual heating effect of adjacent conductors limits the number of conductors allowed to be inserted into it. **Each** conductor that is inserted in a raceway must be counted when determining allowable conduit fill. The number of conductors allowed in a raceway depends not only on the AWG size of the conductor but also on the type of insulation used on the conductor.

Tables #36A (USA only) and #36B (Canada only) provide a listing of the maximum number of conductors of the **same AWG** size allowed in each raceway size.

In order to determine the maximum allowable number of **different AWG** sized conductors in one raceway, tables #37A (USA only) and #37B (Canada only) must be used in conjunction with table #37C.

Allowable Conductors of the Same AWG Size (USA)													
Insulation Type	Size AWG, kcmil	Conduit Trade Size (inches)											
		½	¾	1	1¼	1½	2	2½	3	3½	4	5	6
TW, XHHW	14	9	15	25	44	60	99	142					
	12	7	12	19	35	47	78	111	171				
	10	5	9	15	26	36	60	85	131	176			
	8	2	4	7	12	17	28	40	62	84	108		
RHW and RHH (without outer covering), THW	14	6	10	16	29	40	65	93	143	192			
	12	4	8	13	24	32	53	76	117	157			
	10	4	6	11	19	26	43	61	95	127	163		
	8	1	3	5	10	13	22	32	49	66	85	133	
TW	6	1	2	4	7	10	16	23	36	48	62	97	141
	4	1	1	3	5	7	12	17	27	36	47	73	106
THW	3	1	1	2	4	6	10	15	23	31	40	63	91
	2	1	1	2	4	5	9	13	20	27	34	54	78
	1			1	3	4	6	9	14	19	25	39	57
FEPB (sizes #6 through #2) RHW and RHH (without outer covering)	1/0		1	1	2	3	5	8	12	16	21	33	49
	2/0		1	1	1	3	5	7	10	14	18	29	41
	3/0	1	1	1	1	2	4	6	9	12	15	24	35
	4/0				1	1	3	5	7	10	13	20	29
	250			1	1	1	2	4	6	8	10	16	23
	300			1	1	1	2	3	5	7	9	14	20
	350				1	1	1	3	4	6	8	12	18
	400				1	1	1	2	4	5	7	11	16
	500				1	1	1	1	3	4	6	9	14
	600					1	1	1	3	4	5	7	11
	700					1	1	1	2	3	4	7	10
	750					1	1	1	2	3	4	6	9

Table #36A - Conduit Fill (USA)

Allowable Conductors of the Same AWG Size (Canada)

Insulation-Type	Size AWG, kcmil	Conduit Trade Size (inches)												
		1/2	3/4	1	1 1/4	1 1/2	2	2 1/2	3	3 1/2	4	4 1/2	5	6
TW	14	9	15	25	44	60	99	142	200	200	200	200	200	200
TW75	12	7	12	20	35	47	78	111	171	200	200	200	200	200
	10	5	9	15	26	36	60	85	131	176	200	200	200	200
R90	8	2	4	7	12	17	28	40	62	83	107	134	168	200
Silicone	6	1	1	4	7	10	16	23	36	48	62	78	97	141
(Sizes No. 8 and larger)	4	1	1	3	5	7	12	17	27	36	47	58	73	106
	3	1	1	2	4	6	10	15	23	31	40	50	63	91
	2	1	1	2	4	5	9	13	20	27	34	43	54	78
RW75	1		1	1	3	4	6	9	14	19	25	31	39	57
(XLPE)	0		1	1	2	3	5	8	12	16	21	27	33	49
	00		1	1	1	3	5	7	10	14	18	23	28	41
R90	000			1	1	1	2	4	6	9	12	15	19	24	35
(XLPE)	0000			1	1	1	3	5	7	10	13	16	20	29
	250				1	1	2	4	6	8	10	13	16	23
RW90	300				1	1	2	3	5	7	9	11	14	20
(XLPE)	350				1	1	1	3	4	6	8	10	12	18
	400				1	1	1	2	4	5	7	9	11	16
	500					1	1	1	3	4	6	7	9	14
	600					1	1	1	3	4	5	6	7	11
	700						1	1	2	3	4	5	7	10
	750						1	1	2	3	4	5	6	9
	800						1	1	1	3	4	5	6	9
	900						1	1	1	2	3	4	5	8
	1000						1	1	1	2	3	4	5	7
	1250							1	1	1	2	3	4	6
	1500							1	1	1	1	3	3	5
	1750								1	1	1	2	3	4
	2000								1	1	1	1	2	4

Table #36B - Conduit Fill (Canada)

Size	Types RFH-2, RH, RHH, RHW, SF-2		Types TF, THW, TW		Types TFN, THHN, THWN	
AWG, kcmil	Approx. Diam. in.	Approx. Area sq.in.	Approx. Diam. in.	Approx. Area sq.in.	Approx. Diam. in.	Approx. Area sq.in.
18	.146	.0167	.106	.0088	.089	.0062
16	.158	.0196	.118	.0109	.100	.0079
14	.171	.0230	.131	.0135	.105	.0087
14	*.204	*.0327	**.162	**.0206		
12	.188	.0278	.148	.0172	.122	.0117
12	*.221	*.0384	**.179	**.0252		
10	.242	.0460	.168	.0222	.153	.0184
10			**.199	**.0310		
8	.328	.0845	.245	.0471	.218	.0373
8			**.276	**.0598		
6	.397	.1238	.323	.0819	.257	.0519
4	.452	.1605	.372	.1087	.328	.0845
3	.481	.1817	.401	.1263	.356	.0995
2	.513	.2067	.433	.1473	.388	.1182
1	.588	.2715	.508	.2027	.450	.1590
1/0	.629	.3107	.549	.2367	.491	.1893
2/0	.675	.3578	.595	.2781	.537	.2265
3/0	.727	.4151	.647	.3288	.588	.2715
4/0	.785	.4840	.705	.3904	.646	.3278
250	.868	.5917	.788	.4877	.716	.4026
300	.933	.6837	.843	.5581	.771	.4669
350	.985	.7620	.895	.6291	.822	.5307
400	1.032	.8365	.942	.6969	.869	.5931
500	1.199	.9834	1.029	.8316	.955	.7163
600	1.233	1.1940	1.143	1.0261	1.058	.8791
700	1.304	1.3355	1.214	1.1575	1.129	1.0011
750	1.339	1.4082	1.249	1.2252	1.163	1.0623
800	1.372	1.4784	1.282	1.2908	1.196	1.1234
900	1.435	1.6173	1.345	1.4208	1.259	1.2449
1000	1.494	1.7530	1.404	1.5482	1.317	1.3623
1250	1.676	2.2062	1.577	1.9532		
1500	1.801	2.5475	1.702	2.2751		
1750	1.916	2.8832	1.817	2.5930		
2000	2.021	3.2079	1.922	2.9013		

Note: * - for RHH and RHW only,   ** - for THW only

Table #37A - Dimensions of Conductors (USA)

Dimensions of Insulated Conductors (Canada)

Size AWG, kcmil	* Types RW90, RW75 EP, RW90 EP, RW75, XLPE, RW90 XLPE		** Types TW, TW75, RW75 XLPE, RW90 XLPE, R90 XLPE	
	Diam. in.	Area Sq.in.	Diam. in.	Area Sq.in.
14	0.171	0.0230	0.131	0.0135
12	0.188	0.0278	0.148	0.0172
10	0.242	0.0460	0.168	0.0224
8	0.311	0.0760	0.248	0.0475
6	0.397	0.1238	0.323	0.0819
4	0.452	0.1605	0.372	0.1087
3	0.481	0.1817	0.401	0.1263
2	0.513	0.2067	0.433	0.1473
1	0.588	0.2715	0.508	0.2027
0	0.629	0.3107	0.549	0.2367
00	0.675	0.3578	0.595	0.2781
000	0.727	0.4151	0.647	0.3288
0000	0.785	0.4840	0.705	0.3904
250	0.868	0.5917	0.788	0.4877
300	0.933	0.6837	0.843	0.5581
350	0.985	0.7620	0.895	0.6291
400	1.032	0.8365	0.942	0.6969
500	1.119	0.9834	1.029	0.8316
600	1.233	1.1940	1.143	1.0261
700	1.304	1.3355	1.214	1.1575
750	1.339	1.4082	1.249	1.2252
800	1.372	1.4784	1.282	1.2908
900	1.435	1.6173	1.345	1.4208
1000	1.494	1.7531	1.404	1.5482
1250	1.676	2.2062	1.577	1.9532
1500	1.801	2.5475	1.702	2.2748
1750	1.916	2.8895	1.817	2.5930
2000	2.021	3.2079	1.922	2.9013

Note: * - insulation with a jacket, ** - insulation without a jacket

Table #37B - Dimensions of Conductors (Canada)

Cross-Sectional Area of Conduit and Tubing (USA & Canada)

Size inches	Internal Diameter inches	Cross-Sectional Area - square inches			
		Total 100%	2 Cond. 31%	Over 2 Cond. 40%	1 Cond. 53%
½	0.622	.30	.09	.12	.16
¾	0.824	.53	.16	.21	.28
1	1.049	.86	.27	.34	.46
1¼	1.380	1.50	.47	.60	.80
1½	1.610	2.04	.63	.82	1.08
2	2.067	3.36	1.04	1.34	1.78
2½	2.469	4.79	1.48	1.92	2.54
3	3.068	7.38	2.29	2.95	3.91
3½	3.548	9.90	3.07	3.96	5.25
4	4.026	12.72	3.94	5.09	6.74
5	5.047	20.00	6.20	8.00	10.60
6	6.065	28.89	8.96	11.56	15.31

Table #37C - Cross-sectional Area of Conduit (USA & Canada)

Table #37A (USA) example:

Calculate the minimum size of conduit or tubing required if the following conductors were in one raceway - 2 #14 THWN, 2 #10 RHW, 2 #6 THW and 1 #8 TFN.

Solution: From table #37A obtain the area in sq. in. for each conductor, then add the totals of the areas and refer to table #37C for size of conduit.

Table #37A (USA) example (cont'd):

#14 THWN-0.0087, 2 x .0087 = 0.0174
#10 RHW-0.0460, 2 x .0460 = 0.092
#8 TFN-.0373, 0.0373
#6 THW-0.0819, 2 x .0819 = 0.1638
   Total area = 0.3105 sq. in.

Refer to table #37C and use the column entitled "over 2 conductors - 40%". A 1 inch raceway is required.

Table #37B (Canada) example:

Calculate the minimum size of conduit or tubing required if the following conductors were in one raceway - 2 #14 RW75 XLPE (with a jacket), 2 #10 RW90 XLPE (with a jacket), 2 #6 TW75 (without a jacket) and 1 #8 R90 XLPE (without a jacket).

Solution: From table #37B obtain the area in sq. in. for each conductor and then add the totals of the areas and refer to table #37C for size of conduit.

#14 RW75 XLPE-0.023, 2 x .023 = 0.046
#10 RW90 XLPE-0.046, 2 x .046 = 0.092
#8 R90 XLPE-0.0475, 0.0475
#6 TW75-0.0819, 2 x .0819 = .1638

   Total area = 0.3493 sq. in.

Refer to table #37C and use the column entitled "over 2 conductors - 40%". A 1 ¼ inch raceway is required.

Note: If more than 6 current carrying conductors are inserted into a raceway derating factors may need to be applied. See Section Two (page 95) for derating values.

Conduit Expansion Factors

All materials will expand and contract with changes in temperature. Long raceways when installed in locations where the temperature may vary significantly, such as in outdoor installations, may require expansion joints.

Conduit Expansion Factors (cont'd)

It is recommended that an expansion joint be installed in any conduit where the change in the length of the conduit will exceed 1.77 in. (45 mm). Table #38 lists the coefficients of linear expansion for various conduits.

Table #38 example:

Calculate the change in length of a Type 80 rigid non-metallic conduit (PVC) with a length of 100 feet (30.5 m)

It has a minimum expected temperature of -40°F (-40°C) and a maximum expected temperature of 86°F (30°C).

Imperial solution:
100 x (40 + 86) x 0.000346894 = 4.37 inches

Metric solution:
30.5 x (40 + 30) x 0.052 = 111.02 mm

Coefficient of Linear Expansion		
Material	**Inch per foot per °F**	**mm per m per °C**
Steel Conduit	0.000076	0.0114
Rigid Non-metallic Conduit Schedule 40 (Type RE)	0.000072 - 0.00009	0.0108 - 0.0135
Aluminum Conduit	0.0001467	0.0220
Rigid Non-metallic Conduit Schedule 80 (PVC)	0.000346894	0.0520

Table #38 - Coefficient of Linear Expansion

Flexible Conduit

There are two categories of flexible conduit: metallic and liquid tight (available in both metallic and nonmetallic). There are several advantages to using flexible conduit:

1. Quick and easy to install.
2. No need for special bending equipment or tools.
3. Connection to equipment where vibration is present (motors, compressors etc.).
4. Connection to lighting equipment (T-bar, outdoor luminaire etc.).
5. Connection to portable equipment (welders, heaters, etc.).
6. Liquid tight for connection to equipment in damp locations, corrosive areas or near equipment that may have liquids that splash on to the conduit.
7. Allows for access to or around cumbersome objects.
8. Connections to instruments that are frequently removed for calibration and maintenance.

Manufacturers make a variety of jackets for liquid tight conduits for use in various applications. Metallic flexible conduit is available in both galvanized steel and in aluminum. Sizes available range from 3/8 to 4 in. (10 - 100 mm) in diameter. Examples are shown in illustration #83.

Electrical Nonmetallic Tubing (ENT)

Applications requiring superior flexibility where movement between equipment is present or where vibration and constant flexing are involved may make use of ENT. Applications include process control equipment, laser equipment, computers and peripherals, robotics, office equipment and laboratory equipment. Illustration #84 shows the corrugated appearance of the ENT. ENT is available from 3/8 to 2 in. (10 - 50 mm) in diameter.

Cable Trays

A cable tray is an assembly of sections, fittings, couplers, transition units, etc. that make a complete rigid structure used to support cables and other raceways.

Flexible/ENT/Cable Trays (cont'd)

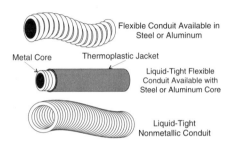

Flexible Conduit Available in Steel or Aluminum

Metal Core — Thermoplastic Jacket

Liquid-Tight Flexible Conduit Available with Steel or Aluminum Core

Liquid-Tight Nonmetallic Conduit

Illustration #83 - Flexible Conduit Examples

Seal-Tight Connector

Illustration #84 - Electrical Nonmetallic Tubing (ENT)

Illustration #85 shows a section of a cable tray system using a ladder type cable tray.

There are four general types of cable trays as shown in illustration #86.

- The ladder cable tray (illustration #86A) is the most common type and is the least expensive, easiest to install and the lightest in weight.

- The ventilated or louvered type cable tray (illustration #86B) has greater mechanical strength but is heavier and costlier.

- The solid type cable tray (illustration #86C) is available in both solid bottom and solid top form and provides additional mechanical protection for cables.

- The tray channel (illustration #86D) provides for wiring to individual branch loads from the main cable tray runs and is available in 3 inch (75 mm) and 4 inch (100 mm) widths and 10 foot (3 m) lengths. Tray channel is also available in both ventilated and solid forms.

Cable Trays (cont'd)

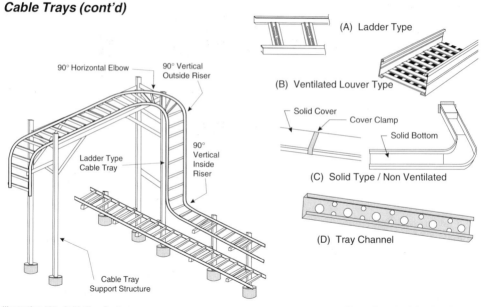

90° Horizontal Elbow

90° Vertical Outside Riser

Ladder Type Cable Tray

90° Vertical Inside Riser

Cable Tray Support Structure

(A) Ladder Type

(B) Ventilated Louver Type

Solid Cover

Cover Clamp

Solid Bottom

(C) Solid Type / Non Ventilated

(D) Tray Channel

Illustration #85 - Cable Tray System

Illustration #86 - Cable Tray Types

Cable Trays (cont'd)

Cable tray is available in the following materials:

a. galvanized steel
b. aluminum
c. PVC coated steel or aluminum
d. stainless steel
e. fiber reinforced plastic

Cable trays are available in standard widths and lengths as shown in table #39.

Cable trays are available in depths ranging from 2 to 6 inches (50 mm - 150 mm).

To determine actual and perceived future mechanical loadings of a cable tray, the following factors should be considered:

a. the weight of present cables
b. the weight of future cables
c. cover plates
d. accessories and fittings
e. auxiliary equipment that may be mounted or attached to the tray
f. snow or ice loading
g. wind shear
h. drilling of holes in cable tray
i. weight of personnel walking on tray

Cable Tray Dimensions				
Type	**Standard Widths**		**Standard Lengths**	
	inches	mm	feet	m
Ladder	6	150		
	12	300	10	3
	18	450	& 20	& 6
	24	600		
	30	750		
	36	900		
Ventilated/ Louvered	6	150		
	12	300	10	3
	18	450	& 20	& 6
	24	600		
Solid/ Unventilated	6	150		
	12	300	10	3
	18	450	&20	& 6
	24	600		

Table #39 - Cable Tray Standard Dimensions

Cable Trays (cont'd)

Cable Tray Loading Standards						
Class of Tray	Support Spacing (feet) and Design Load (pounds per foot)					
	5	6.5	8	10	13	16	20
A	66.2	41.46	30.1	24.7			
C_1	173.2	109.7	79.57	64.9			
D_1				119.7	75.6	54.8	44.8
E				200	126.4	91.6	74.9
Class of Tray	Support Spacing (m) and Design Load (kg per m)					
	1.5	2	2.5	3	4	5	6
A	99	62	45	37			
C_1	259	164	119	97			
D_1				179	113	82	67
E				299	189	137	112

Table #40 - Cable Tray Loading Standards

Each of the cable tray classes has a mechanical loading rating. Mechanical loading standards are given in table #40.

Both the NEC and the CEC place restrictions on the number of cables or conductors that are permissible within a cable tray.

If the values, as per the NEC and CEC, are exceeded, then the current carrying capacity of the individual conductors must be derated.

See NEC Article 318 or the CEC Rule 12-2212 for specific derating values and spacings of conductors in cable trays.

Cables/Conductors in Cable Trays

The NEC and CEC provide clear guidelines as to which cables and conductors are allowed in cable trays. The following cables and conductors may be placed within a cable tray assembly:

- mineral-insulated metal-sheathed cable
- armoured cable
- aluminum-sheathed cable
- tray cable (see illustration #87)
- nonmetallic-sheathed cable (USA only)

Support Clamps and Accessories

Support brackets are available in prefabricated form or they may be field constructed. Illustration #88 shows both prefabricated and field assembled cable tray supports. Cable clamps and vertical cable supports are shown in illustration #89.

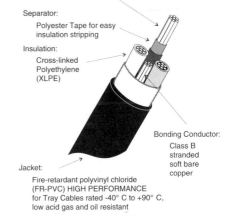

Conductors:
Class B soft copper, stranded to reduce cable diameter and weight
sizes 14 - 10 AWG are compressed to reduce weight
sizes 8 AWG to 1000 MCM/kcmil are compact

Separator:
Polyester Tape for easy insulation stripping

Insulation:
Cross-linked Polyethylene (XLPE)

Bonding Conductor:
Class B stranded soft bare copper

Jacket:
Fire-retardant polyvinyl chloride (FR-PVC) HIGH PERFORMANCE for Tray Cables rated -40° C to +90° C, low acid gas and oil resistant

Illustration #87 - Tray Cable Construction

Support Clamps/Accessories (cont'd)

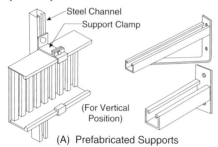

(A) Prefabricated Supports

Steel Channel
Support Clamp
(For Vertical Position)

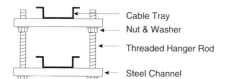

(B) Field Assembled Trapeze Hanger

Cable Tray
Nut & Washer
Threaded Hanger Rod
Steel Channel

Illustration #88 - Cable Tray Supports

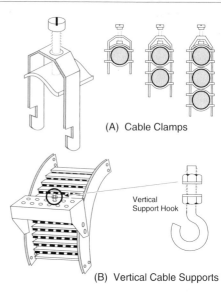

(A) Cable Clamps

(B) Vertical Cable Supports

Vertical Support Hook

Illustration #89 - Cable Supports

Cable Tray Grounding

Grounding of the cable tray or carrying a grounding/bonding conductor may be necessary if the cable tray does not meet Code requirements (ie: largest overcurrent device protecting cables in tray is greater than metal trays rating, inadequate contact with earth through structural members or the cable tray is nonmetallic).

Illustration #90 provides two examples of ground clamps attached to a cable tray.

Cable Rollers and Pulleys

To facilitate cable installation, it may be necessary to use cable rollers and pulleys as shown in illustration #91. A mechanical cable tugger is commonly employed in conjunction with the equipment shown in illustration #91.

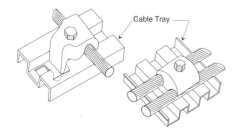

Illustration #90 - Grounding Clamps

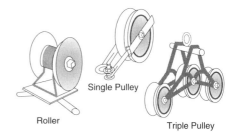

Illustration #91 - Cable Rollers and Pulleys

Busways

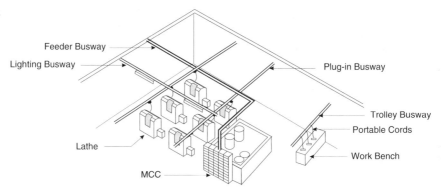

Illustration #92 - Busway System

A busway is a prefabricated assembly of copper or aluminum busbars separated from each other by thin insulators and housed within a rigid metallic structure. Busways are designed to accommodate accessories such as fused and unfused switches, circuit breakers, termination/transition points and fittings.

Illustration #92 provides an example of some of the parts of a busway system.

Busway systems may be divided into two categories, and within each category there are numerous types of busways:

Busways (cont'd)

Busways 600V and under

a. Feeder busway for low-impedance bulk transmission of power. Ratings available are from 600 A to 5000 A (see illustration #93A).

b. Plug-in busway for connection to or re-arrangement of loads. Ratings available are from 100 A to 4000 A (see illustration #93B).

c. Lighting busway to provide power to individual luminaires and also provide mechanical support for the mounting of the luminaire. Rated 60 A (see illustration #93C).

d. Trolley busway is constructed to receive stationary or movable take-off devices and is typically used on production lines where the operator moves back and forth. The trolley busway may also be used to supply motor loads on travelling cranes. Ratings are usually up to 100 A (see illustration #93D).

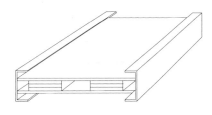

Illustration #93A - Feeder Busway

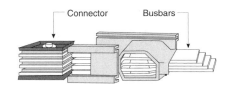

Illustration #93B - Plug-in Busway

Busways (cont'd):

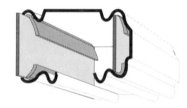

Illustration #93C - Lighting Busway

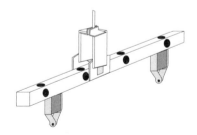

Illustration #93D - Trolley Busway

Busways 600V and over

(usually referred to as metal-enclosed bus)

a. Isolated phase bus is a utility busway used in power generating stations.
b. Segregated phase bus is a utility busway used in power generating stations.
c. Non-segregated phase bus is an industrial busway used for non-utility applications such as connection of high voltage transformers to their switchgear. This busway is available in voltage ratings of 4.16 kV, 13.8 kV, 23 kV, and 34.5 kV. Current ratings depend on the voltage but are available from 1200 A to 3000 A. Non-segregated phase bus are totally enclosed in steel, stainless steel or aluminum. The aluminum or copper busbars are supported on glass, polyester or porcelain supports.

Busway Uses

Busways are particularly advantageous when numerous taps to loads are required or

Busway Uses (cont'd)

when large ampacities need to be transported from the source transformer(s) to the distribution center(s).

Busways are easy to assemble, dismantle, or add to, thus allowing for changes to the electrical distribution system. Busways are typically used in automotive assembly line plants, manufacturing plants, feeder risers in high rise buildings and assembly production buildings.

Both the NEC Article 364 and the CEC Rule 12-2000 place limitations on the use of busways. Derating factors, permissible locations, supports, overcurrent protection, grounding and manufacturers' markings are dealt with in the NEC and the CEC. Since there are such variances in the types and applications of busways it is recommended that the respective manufacturers of busways be consulted for specific information.

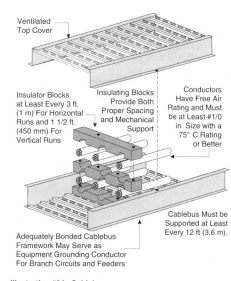

Ventilated Top Cover

Insulator Blocks at Least Every 3 ft. (1 m) For Horizontal Runs and 1 1/2 ft. (450 mm) For Vertical Runs

Insulating Blocks Provide Both Proper Spacing and Mechanical Support

Conductors Have Free Air Rating and Must be at Least #1/0 in Size with a 75° C Rating or Better

Cablebus Must be Supported at Least Every 12 ft (3.6 m).

Adequately Bonded Cablebus Framework May Serve as Equipment Grounding Conductor For Branch Circuits and Feeders

Illustration #94 - Cablebus

Cablebus

Cablebus is an approved assembly of insulated thermoplastic or thermoset conductors, mounted in a spaced relationship on nonmetallic support blocks, and housed in a ventilated metal enclosure. Illustration #94 provides a view of a cablebus and its Code requirements. Cablebus is commonly used when large amounts of current need to be transported over relatively short distances (ie: up to several thousand feet).

Cablebus offers a safe, reliable and economical alternative to conventional feeder wiring such as busways or conduit and wire. Pulp and paper plants are now using more cablebus due to its ability to tolerate both corrosive and wet environments. Cablebus is also popular in the power generating industry.

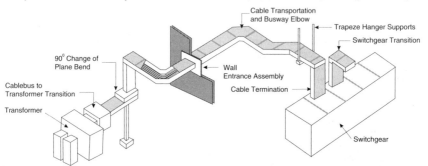

Illustration #95 - Cablebus Layout

Cablebus (cont'd)

Cablebus is available in voltages up to and including 69 kV. Ampacities vary from 600 A to 5000 A. Cablebus is designed to withstand severe fault currents up to and including 100,000 A asymmetrical. Using non-magnetic metal (aluminum) enclosures results in lower hysteresis losses and higher corrosion resistance. Using aluminum enclosures requires less maintenance than conventional painted steel enclosures. Illustration #95 shows a layout of cablebus and the necessary hardware used to connect and assemble cablebus to the electrical distribution equipment.

Cablebus Parallel Arrangement

Conductors are either Class B stranded soft annealed copper or aluminum alloy. They are usually 500 MCM (kcmil), 750 MCM (kcmil) or 1000 MCM (kcmil) with a cross-linked polyethylene insulation.

To achieve higher amperage ratings, the individual conductors are arranged in an approved parallel arrangement, as shown in illustration #96. The arrangement of these parallel conductors is critical in order that each phase group has the same impedance and that each conductor within each phase group has the same impedance. Unequal impedances result in voltage unbalances and overheating of some of the conductors.

3 Conductor

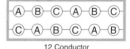

12 Conductor

6 Conductor

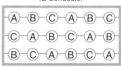

18 Conductor

Illustration #96 - Cablebus Parallel Arrangement

Cablebus Installation

The basic installation of cablebus is similar to that of cabletrays; therefore, the same tools and equipment may be used. However, cablebus can be shipped in a pre-assembled state if the lengths are short.

Cablebus is usually field assembled from shipped manufactured parts and assembled according to these steps:

1. Construct the support structures for the cablebus enclosure and prepare openings in walls and floor through which the cablebus must be routed.
2. Install the metal enclosure of the cablebus.
3. Pull the conductors into the metal assembly after the basic framework is in place.
4. Install the mounting blocks upon which the conductors are placed, and then secure the conductors to these blocks.
5. Install the top covers.

Note: NEC Article 365 and the manufacturers' recommended procedures and specifications should be followed when installing and choosing cablebus.

Wireways

A wireway is a metal trough with a removable or hinged cover, so designed that once the wireway has been completely assembled conductors may be laid within it. Typical sizes of a wireway are 2 1/2, 4, 6 and 8 inches (63.5, 101.6, 152.4, 203.2 mm) square. The main advantage and application of wireways involves the need to have access to branching circuits or where many conductors must be located in one run. Wireways permit access to conductors at any point and connection to wireways may be made with conduit (rigid or flexible), armoured cable, EMT, surface raceways, cabletrays, and even with approved cords.

Wireways (cont'd)

Illustration #97 shows a wireway arrangement complete with conductors and associated approved wiring.

Wireways may only be installed for exposed work and preferably indoors in a dry location. However, wireways are made and approved for damp or wet locations.

Both the NEC Article 362 and the CEC Rule 12-2100 place restrictions on the use of wireways regarding their location, type and size of conductors, grounding, mounting, support and splicing of conductors. Wireways may be coupled together, tees may be made, elbows may be installed and transition plates may be used. Wireways offer a flexible means of wiring when multiple access points and conductors are required within one assembly.

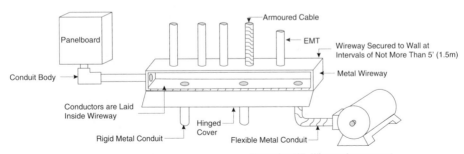

Illustration #97 - Wireway Arrangement

Underfloor Raceways

Underfloor raceways are a means of providing electrical power, signal and communication wiring to large offices, retailing and commercial areas. Underfloor raceways are available in both metallic and nonmetallic forms and must be installed either within the floor slab or flush with the floor surface.

NEC Article 354 or CEC Rule 12-1700 along with the manufacturers' recommendations should be referred to when planning and installing this type of raceway.

Illustration #98 shows an exposed view of a two-level underfloor raceway system.

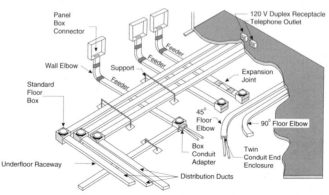

Illustration #98 - Underfloor Raceway System

Underfloor Raceways (cont'd)

Underfloor raceways are available in various sizes and in single level and two level designs (see illustration #99). The raceway must be covered with concrete or wood to a specified depth depending upon the width of the raceway. Consult local code authorities and the manufacturer for recommended minimum values.

Underfloor duct systems are very expensive and the justification for their use is diminishing as newer or enhanced alternative wiring methods are introduced.

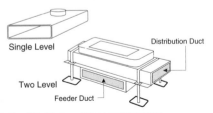

Illustration #99 - Single and Two Level Raceways

The use of underfloor raceways may increase electrical construction costs by as much as 50% making this wiring method cost prohibitive. The following list is a basis for determining where underfloor raceways may be appropriately used:

a. For large open floor areas with electrical requirement for numerous outlets positioned some distance from walls and partitions.

b. When use of underfloor carpet wiring systems is not possible or acceptable.

c. When wiring from the ceiling, via pack poles, etc. is not possible or acceptable.

d. When the floor layout plan is subject to change, requiring the availablity of wiring within the floor in new locations.

e. When access from the ceiling space above is not possible or practical.

Use of underfloor raceways may be justifiable from a cost standpoint in some prestige office buildings, museums, galleries, high-cost merchandising areas and even in some selected areas of industrial buildings.

Cellular Floor Raceways

Cellular floors are an assembly of cellular metal or precast concrete floor members with hollow cells or spaces suitable for use as raceways. Illustration #100 shows a typical layout of a cellular floor raceway. NEC Articles 356 and 358, or CEC Rule 12-1800, and the raceway manufacturers' recommendations should be consulted.

The cost, complexity and coordination required between the structural and electrical arrangements needed to set up the wiring must be considered, along with the flexibility required to access the wiring after installation. All factors listed for underfloor raceways must also be considered before adopting this means of wiring.

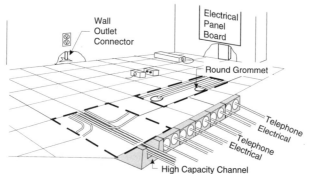

Illustration #100 - Cellular Floor Raceway

Surface Raceways

A surface raceway is a metallic or non-metallic channel designed to hold conductors for power, communications, or signalling. NEC Article 352, or CEC Rule 12-1600 and the manufacturers' specifications should be consulted prior to designing or installing surface raceways. The decision to use surface raceways depends upon a number of factors:

1. The architecture or structure of the building space does not allow for the electrical wiring to be recessed.
2. The economics of surface raceways is significantly better than using other wiring methods.
3. The need for multiple outlets/circuits along wall spaces.
4. Access to the wiring or the outlet is required.
5. To provide new wiring to existing buildings with minimum disruption to the walls, ceilings and floors.
6. Trying to achieve a certain appearance and facilitate the electrical wiring.
7. Interfacing with flat conductor cable (undercarpet wiring).

Manufacturers have made available a wide selection of both metallic and nonmetallic surface raceways, fittings and accessories. Illustration #101 provides some examples of the products that are in use for surface raceways. Surface metal or nonmetal raceways should be installed in dry, non-hazardous, non-corrosive locations, and should contain conductors operating below 300 V. This type of a raceway should be installed where it may be exposed and free from physical damage. Various tools are used for installing surface raceways including benders, cutting boxes, wire pulleys, fish tape leaders, canopy cutters, and base and cover shears.

Surface Raceways (cont'd)

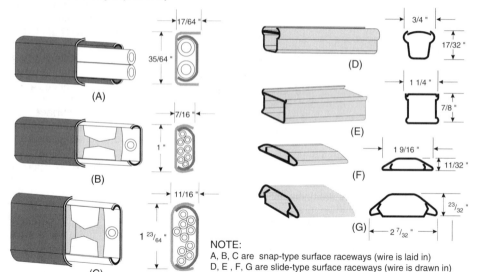

NOTE:
A, B, C are snap-type surface raceways (wire is laid in)
D, E , F, G are slide-type surface raceways (wire is drawn in)

Illustration #101A-G - Surface Raceways

Surface Raceways (cont'd)

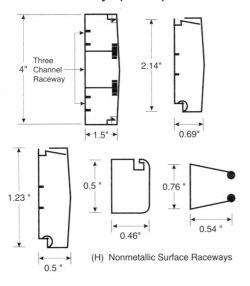

Three Channel Raceway

4"

1.5"

2.14"

0.69"

1.23"

0.5"

0.5"

0.46"

0.76"

0.54"

(H) Nonmetallic Surface Raceways

(I) Nonmetallic Surface Raceways c/w Conductors

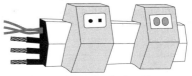

(J) Outlets Mounted on Surface Raceway

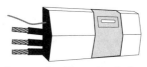

(K) Computer Port Mounted on Surface Raceway

Illustration #101H-K - Surface Raceways

Raised Floors

Raised floors or full access floors provide easy access to underfloor plenums. This type of a system, as shown in illustration #102, grew from the needs imposed by huge data processing systems. Access to various pieces of computer equipment needed to be gained from the floor, and future cabling requirements meant that the wiring system had to be flexible and accessible.

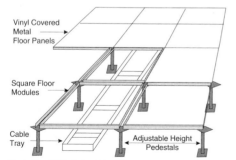

Vinyl Covered
Metal
Floor Panels

Square Floor
Modules

Cable
Tray

Adjustable Height
Pedestals

Illustration #102 - Raised Floor System

The raised floor system consists of die cast aluminum panels supported on a network of adjustable steel or aluminum pedestals. Floor panels are available in sizes ranging from 18 inches x 18 inches (450 x 450 mm) to 36 inches x 36 inches (900 x 900 mm) and in depths from 12 in. to 36 in. (300 to 900 mm).

Raised floors are not considered to be raceways but rather plenums, and, therefore, cables and conductors must be placed within approved raceways which are then placed within the plenum space. A cable tray is commonly used as an approved raceway because it allows for changes in the wiring system after its initial installation. Fire Code and National Building Code regulations must be consulted, and cables and conductors must be fire code rated. Often the floor plenum is provided with fireproofing by using dry chemical or oxygen evacuation extinguishing. NEC and CEC require raised floor pedestals to be bonded.

SECTION FOUR

WIRING DEVICES

Introduction

In this section the term *wiring devices* will be limited to plugs, receptacles and switches.

A *plug* may be defined as a wiring device that is either designed to be inserted (male) into the slots of a receptacle, or designed to receive (female) the prongs of a male plug as shown in illustration #103.

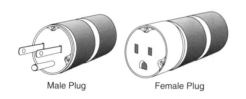

Male Plug   Female Plug

Illustration #103 - Male and Female Plugs

A *receptacle* may be defined as one or more female contact devices, on the same assembly, installed at an outlet for the connection of one or more attachment plugs (male).

A *switch* may be defined as a device for making, breaking, or changing circuit connections.

Plugs are either cord or cable connected devices and do not require an enclosure to house them. Enclosures are required for receptacles and switches, and these enclosures must be rated for the environment and atmosphere in which they are placed. Both NEMA (USA) and EEMAC (Canada) have enclosure designations to indicate the environmental conditions for which the enclosures are suitable. Refer to illustration #104 for enclosure types and their designations as specified by NEMA and EEMAC.

Enclosure Types

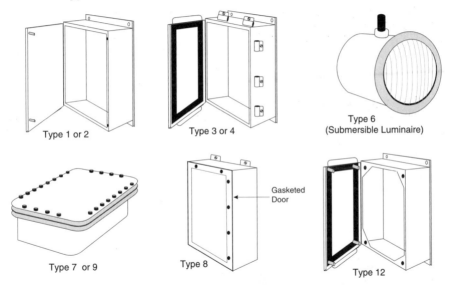

Type 1 or 2

Type 3 or 4

Type 6
(Submersible Luminaire)

Type 7 or 9

Type 8

Gasketed
Door

Type 12

Illustration #104 - Enclosure Types

Enclosure Designations

Type 1: Intended for indoor use and may either be ventilated or nonventilated. Serves as a protection against falling dust but is not dust tight. It is primarily designed to protect against contact with the enclosed equipment in locations where unusual service conditions do not exist (general purpose).

Type 2: Intended for indoor use and may either be ventilated or nonventilated. Serves to provide a degree of protection against limited falling water and dirt by using drip shields or other means. It is suitable for application where external condensation may be severe but does not protect against internal condensation.

Type 3: Intended for outdoor use and may either be ventilated or nonventilated. Serves to provide protection against wind-blown dust, rain, sleet, ice and snow. May also be used indoors where falling or light splashing of water may be present. Type 3 enclosures do not protect against internal icing or condensation.

Type 4: Intended for either indoor or outdoor use when exposed to wind-blown dust, rain, splashing water, and hose-directed streams of water directed at any angle. These enclosures do not protect against internal icing or condensation and are not approved for water submersion.

Type 5: Intended for indoor use to protect against dust and falling dirt. These enclosures do not protect against internal condensation.

Type 6: Intended for indoor or outdoor use to protect equipment that may be submersed in water. There are two ratings for Type 6 enclosures: Type 6 and Type 6P. Type 6 provides protection against the entry of water during occasional temporary submersion at a limited depth. Type 6P provides protection against the entry of water during prolonged submersion at a limited depth.

Enclosure Designations (cont'd)

Type 7: Intended for indoor use in locations classified by the NEC or CEC as Class 1, Groups A,B,C or D (hazardous locations). These enclosures are designed to withstand pressure resulting from an internal explosion of a specified gas. The atmosphere surrounding the enclosure must not ignite. Enclosures containing heat generating devices, such as lamps, must not allow external surfaces to reach temperatures capable of igniting explosive gas-air mixtures in the surrounding atmosphere.

Type 8: Intended for indoor or outdoor use in locations classified by the NEC or CEC as Class 1, Groups A, B, C or D. Devices located within Type 8 enclosures are so arranged that all arcing contacts and connections are immersed in oil.

Arcing is suppressed by the oil to prevent internal ignition. Enclosed heat producing devices, such as lamps, will not cause external surfaces to reach temperatures capable of igniting explosive gas-air mixtures in the surrounding atmosphere.

Type 9: Intended for indoor use in locations classified by the NEC or the CEC as Class II, groups E, F or G. Type 9 enclosures must be capable of preventing the entrance of dust. Enclosed heat generating devices, such as lamps, must not cause external surfaces to reach temperatures capable of igniting or discoloring accumulated dust on the enclosure or of igniting dust-air mixtures in the surrounding atmosphere.

Type 10: Intended to meet the requirements for use of electricity in mines.

Enclosure Designations (cont'd)

Type 11: Intended for indoor use to provide protection by means of oil immersion to enclosed equipment against the corrosive effects of liquids and gases.

Type 12: Intended for indoor use to provide protection against dust, falling dirt and dripping water. Suitable for application to machine tools and other industrial processing machines in locations where oil, coolants, water, filings, dust, or lint might be present. These enclosures do not protect against internal condensation.

Type 13: Intended for indoor use to provide protection against dust, spraying or splashing of water, oil and noncorrosive coolants. These enclosures do not protect against internal condensation.

Boxes

Wiring devices such as receptacles and switches are mounted on or within wiring boxes. Several types are shown in illustration #105.

These boxes have enclosure designations as mentioned in the previous subsection, and have a capacity rating (volume) for holding of wires, terminations, fittings and wiring devices. NEC Article 370, or CEC Rule 12-3038 should be referred to when determining the number of conductors and the size of allowable conductors within a box or enclosure. Tables #41A and #41B apply to USA installations, and tables #42A, #42B and #42C apply to Canadian installations when determining the maximum number of conductors permitted in a box or enclosure.

Boxes (cont'd)

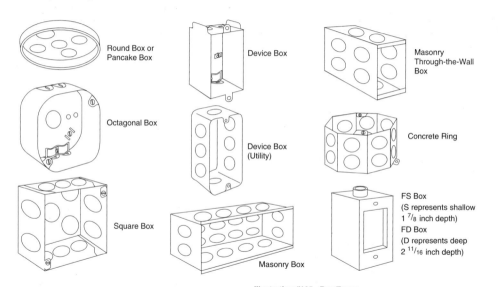

Round Box or Pancake Box

Device Box

Masonry Through-the-Wall Box

Octagonal Box

Device Box (Utility)

Concrete Ring

Square Box

Masonry Box

FS Box
(S represents shallow 1 7/8 inch depth)
FD Box
(D represents deep 2 11/16 inch depth)

Illustration #105 - Box Types

Boxes (cont'd)

Box Size and Conductors (USA)

Type	Box dimensions (inches) or Trade size	Cubic inch capacity	Maximum number of Conductors size AWG						
			18	16	14	12	10	8	6
Round or Octagonal	4 x 1 1/4	12.5	8	7	6	5	5	4	2
	4 x 1 1/2	15.5	10	8	7	6	6	5	3
	4 x 2 1/8	21.5	14	12	10	9	8	7	4
Square	4 x 1 1/4	18.0	12	10	9	8	7	6	3
	4 x 1 1/2	21.0	14	12	10	9	8	7	4
	4 x 2 1/8	30.3	20	17	15	13	12	10	6
	4 11/16 x 1 1/4	25.5	17	14	12	11	10	8	5
	4 11/16 x 1 1/2	29.5	19	16	14	13	11	9	5
	4 11/16 x 2 1/8	42.0	28	24	21	18	16	14	8
Device	3 x 2 x 1 1/2	7.5	5	4	3	3	3	2	1
	3 x 2 x 2	10.0	6	5	5	4	4	3	2
	3 x 2 x 2 1/4	10.5	7	6	5	4	4	3	2
	3 x 2 x 2 1/2	12.5	8	7	6	5	5	4	2
	3 x 2 x 2 3/4	14.0	9	8	7	6	5	4	2
	3 x 2 x 3 1/2	18.0	12	10	9	8	7	6	3
	4 x 2 1/8 x 1 1/2	10.3	6	5	5	4	4	3	2
	4 x 2 1/8 x 1 7/8	13.0	8	7	6	5	5	4	2
	4 x 2 1/8 x 2 1/8	14.5	9	8	7	6	5	4	2
Masonry Box/Gang	3 3/4 x 2 x 2 1/2	14.0	9	8	7	6	5	4	2
	3 3/4 x 2 x 3 1/2	21.0	14	12	10	9	8	7	4
FS—Single/Gang	1 3/4 deep	13.5	9	7	6	6	5	4	2
FS—Multi/Gang	1 3/4 deep	18.0	12	10	9	8	7	6	3
FD—Single/Gang	2 3/8 deep	18.0	12	10	9	8	7	6	3
FD—Multi/Gang	2 3/8 deep	24.0	16	13	12	10	9	8	4

Table #41A - Boxes (USA)

Boxes (cont'd)

Note: Table #41A shall be modified as follows:

1. For each fixture stud, cable clamp, or hickey** contained in the box reduce the allowable number of conductors by one.
2. For each mounting yoke or strap containing one or more wiring devices reduce the allowable number of conductors by two.
3. For each set of grounding conductors entering the box [see NEC 370-6(a)1], reduce the allowable number of conductors by one.
4. Conductors running through the box count as one, and each conductor entering but not going through the box counts as one.

** Hickey: A threaded coupling between an electrical fixture and an outlet box.

Volume Required per Conductor	
Size of Conductor (AWG)	Free space within Box for each Conductor (cubic inches)
No. 18	1.5
No. 16	1.75
No. 14	2.00
No. 12	2.25
No. 10	2.5
No. 8	3.0
No. 6	5.0

Table #41B - Box Volume (USA)

Boxes (cont'd)

Box Size and Conductors (Canada)							
Type	Box dimensions (inches) Trade size	Cubic inch capacity	Maximum number of Conductors size AWG				
			14	12	10	8	6
Octagonal	4 x 1 1/2	15	10	8	6	5	3
	4 x 2 1/8	21	14	12	9	7	4
Square	4 x 1 1/2	21	14	12	9	7	4
	4 x 2 1/8	30	20	17	13	10	6
	4 11/16 x 1 1/2	30	20	17	13	10	6
	4 11/16 x 2 1/8	42	28	24	18	15	9
Round	4 x 1/2	5	3	2	2	1	1
Device	3 x 2 x 1 1/2	8	5	4	3	2	1
	3 x 2 x 2	10	6	5	4	3	2
	3 x 2 x 2 1/4	10	6	5	4	3	2
	3 x 2 x 2 1/2	12.5	8	7	5	4	2
	3 x 2 x 3	15	10	8	6	5	3
	4 x 2 x 1 1/2	9	6	5	4	3	2
	4 x 2 1/8 x 1 7/8	14	9	8	6	5	3
	4 x 2 3/8 x 1 7/8	16	10	9	7	5	3

Table #42A - Boxes (Canada)

Boxes (cont'd)

	Box dimensions (inches) Trade size	Cubic inch capacity	Maximum number of Conductors size AWG				
Type			**14**	**12**	**10**	**8**	**6**
Masonry	3¾ x 2 x 2½	14/gang	9	8	6	5	3
	3¾ x 2 x 3½	21/gang	14	12	9	7	4
	4 x 2¼ x 2⅜	20.25/gang	13	11	9	7	4
	4 x 2¼ x 3⅜	22.25/gang	14	12	9	8	4
Through Box	3¾ x 2	6/inch	4	3	2	2	1
Concrete Ring	4	12/inch	8	6	5	4	2
FS	1 Gang	14	9	8	6	5	3
	1 Gang Tandem	34	22	19	15	12	7
	2 Gang	26	17	14	11	9	5
	3 Gang	41	27	23	18	14	9
	4 Gang	56	37	32	24	20	12
FD	1 Gang	22.5	15	12	10	8	5
	2 Gang	41	27	23	18	14	9
	3 Gang	60	40	34	26	21	13
	4 Gang	85	56	48	37	30	18

Table #42A (cont'd) - Boxes (Canada)

Box Size and Conductors (Canada)

Boxes (cont'd)

Note: Table #42A shall be modified as follows:

1. For each fixture stud and hickey**, reduce the number of wires by one.
2. For every pair of two or three wire connectors with insulating caps, reduce the number of wires by one and reduce by two wires for four or five wire connectors, etc.
3. For each mounting yoke or strap containing a wiring device, reduce the number of wires by two.
4. For boxes containing wiring devices that have a depth dimension greater than one inch, multiply the depth dimension by 5. This value is the cubic inch volume that the device occupies.

** Hickey: A threaded coupling between an electrical fixture and an outlet box.

5. Conductors passing through a box count as one conductor. Each conductor that either enters or leaves a box is counted as one conductor, and a conductor of which no part leaves the box shall not be counted.

Volume Required per Conductor	
Size of Conductor (AWG)	Useable space required for each insulated Conductor (cubic inches)
No. 14	1.5
No. 12	1.75
No. 10	2.25
No. 8	2.75
No. 6	4.5

Table #42B - Box Volume (Canada)

Box and Conductor Table Examples

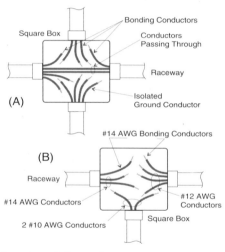

(A)

(B)

Illustration #106 - Box and Conductor Examples

Example 1 (table #41 - USA):

Determine the minimum size of a square box that contains two duplex receptacles mounted on straps and conductors as shown in illustration #106A. Note, all the conductors are #14 gage.

Solution:

1. 4 wires for the two duplex receptacles
2. 2 wires for the conductors passing through
3. 7 wires for the conductors entering the box
4. 1 conductor for the bonding conductors and 1 conductor for the isolated grounding conductor

Total wires = 15, therefore a 4 x 2⅛ inch box is needed.

Note: No deduction is made for conductors that do not enter or leave the box.

Box & Conductor Examples (cont'd)
Example 2 (table #41 - USA):
Determine the minimum size of a square box that contains two switches mounted on straps and conductors as shown in illustration #106B (Note the conductors have different gage sizes).

Solution:

Use table #41B to determine the total volume occupied by the devices and conductors. After the volume has been calculated and expressed in cubic inches, use table #41A to choose a box with sufficient volume.

1. 4 x 2.5 (volume of switches based on largest conductor in box)
2. 2 x 2 (#14 AWG conductors)
3. 3 x 2.25 (#12 AWG conductors)
4. 2 x 2.5 (#10 AWG conductors)
5. 1 x 2 (#14 AWG bonding conductors)

Total volume required = 27.75 cu. in.

Referring to table #41A, a 4$\frac{11}{16}$ x 1$\frac{1}{2}$ inch box is required.

Example 3 (table #42 - Canada):
Refer to Example 1 and illustration #106A, and then add 3 two-wire insulated wire connectors.

Solution:

1. 4 wires for the two wiring devices
2. 2 wires for the conductors passing through
3. 11 wires for the conductors entering the box
4. 2 conductors for the 3 two-wire insulated wire connectors (one conductor for each pair or portion thereof)

Total wires = 19.

Therefore from table #42A, a 4 x 2$\frac{1}{8}$ inch box is required.

Box & Conductor Examples (cont'd)
Example 4 (table #42 - Canada)
Refer to Example 2 and illustration #106B. The depth of each wiring device used in the box is 1½ inches.

Solution:

First determine the cubic inch capacity of all devices and conductors that are within the box (refer to table #42B). After the volume has been determined and expressed in cubic inches, refer to table #42A and choose a box that has sufficient volume.

1. 2 x 5 x 1 1/2 (volume requirement of the two wiring devices)
2. 2 x 1.5 (2 #14 AWG conductors)
3. 3 x 1.75 (3 #12 AWG conductors)
4. 2 x 2.25 (2 #10 AWG conductors)
5. 2 x 1.5 (2 #14 AWG bonding conductors)

Total volume required = 30.75 cu.in.

Referring to table #42A, a 4¹¹⁄₁₆ x 2⅛ inch box is required.

Receptacles

Receptacles are available in a wide variety of voltages (600 V or less) and amperage ratings (10 to 400 A).

Note: Most receptacles are of the grounding type, and all new installations are required by Code to be grounded. Older nongrounded receptacles must not be replaced with the new grounding types unless provision has been made to bring a grounding/bonding conductor to the receptacle.

Table #43 provides a listing of the nonlocking configurations of general-purpose receptacles. These configurations conform with NEMA and EEMAC standards.

Nonlocking Receptacles

	Nonlocking Receptacle Configurations										
Ampere Rating	2-pole 2-wire		2-pole 3-wire grounding				3-pole 3-wire		3-pole 4-wire grounding		4-pole 4-wire
	125 V	250 V	125 V	250 V	277 V AC	347 V AC	125 / 250 V	3 Φ 250 V	125 / 250 V	3 Φ 250 V	3 Φ 208 Y / 120 V
15 A	⊡		⊡	⊡	⊡	⊡		⊡	⊡	⊡	⊡
20 A		⊡	⊡	⊡	⊡	⊡	⊡	⊡	⊡	⊡	⊡
30 A		⊡	⊡	⊡	⊡	⊡	⊡	⊡	⊡	⊡	⊡
50 A			⊡	⊡	⊡	⊡	⊡	⊡	⊡	⊡	⊡
60 A	NOTE: Plugs are mirror images of the receptacles								⊡	⊡	⊡

Table #43 - Nonlocking Receptacles

General Purpose Receptacles

General purpose receptacles are available in the following forms or grades:

Residential Grade: This receptacle is intended for light duty operations and in areas where mechanical strength requirements are very light. This receptacle is most commonly used for residential applications in areas where light operational duty is normal. ***For heavier duty, such as in the kitchen area, a higher grade of receptacle is recommended***.

Specification Grade: This receptacle varies in its ruggedness, dependability and duty rating. Manufacturers have a great variance in this grade and therefore, each receptacle must be carefully analyzed as to its suitability for the intended application. ***This receptacle is commonly found in offices, commercial and retail outlets and in industrial shops***.

Hospital Grade: This receptacle must meet ANSI/UL 498-1980 standards. These receptacles have much higher mechanical and electrical strength characteristics. Among the criteria are the stringent minimum retention and ground resistance specifications. ***Hospital grade receptacles are identified with a green dot on the face of the receptacle***.

Isolated Ground: This receptacle comes in various grades and is the same as a conventional receptacle except for an additional feature, which is an insulating barrier isolating the ground contacts from the mounting strap (see illustration #107B). The grounding screw is connected directly to the grounding contacts, and no electrical continuity exists between the mounting strap and the bonding conductor. The insulated bonding/grounding conductor does not terminate at any other point than the equipment grounding conductor terminal of the derived system or service and the intended isolated ground receptacle.

General Purpose Receptacles (cont'd)

Isolated Ground (cont'd): The purpose of isolating the ground is to keep electrical noise that may be impressed on the ground to a minimum. Sensitive electronic equipment could malfunction if subjected to transient electrical signals.

Isolated ground receptacles are identified by an orange triangle on the face of the receptacle.

Surge Suppression Receptacle: This receptacle has built-in electronic circuitry to absorb and dissipate transient surge voltages. The surge suppression receptacle clamps voltage surges to acceptable values. Computers and other sensitive electronic equipment are protected against transient voltages that could disable or damage them. *Surge suppression receptacles are identified with a distinctive marking and/or coloration.*

Ground Fault Circuit Interrupter (GFCI): These receptacles are designed to protect people from the effects of line - to - ground electrical current leakage. Illustration #108A shows how the receptacle is connected and #108B how it operates. GFCI will respond to leakage values in excess of 4 to 6 milliamps and interrupt the electrical circuit if values above these are detected. The greater the leakage current, the faster the interruption time. GFCI devices do not protect against line to neutral shock hazards.

Split - Bus Receptacles: 15 A or 20 A, 125 V duplex receptacles are available with break-off links or removable jumpers so that one or both sides of the receptacle can be separated as shown in illustration #109. The purpose of separating one or more of the links from each other is to connect each part of the receptacle to a different circuit or to control one half of the receptacle from a switch.

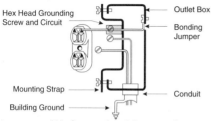

(A) Conventional Receptacle

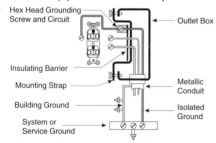

(B) Isolated Ground Receptacle

Illustration #107 - Conventional Isolated Ground

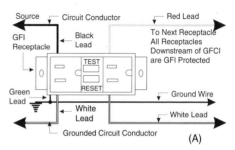

(A)

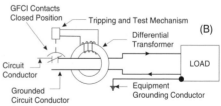

(B)

If the current in the circuit conductor is not equal to the current in the grounded circuit conductor, the differential transformer will produce an MMF thereby causing the tripping mechanism to be released.

Illustration #108 - GFCI Receptacle

General Purpose Receptacles (cont'd)

(A) Duplex Receptacle

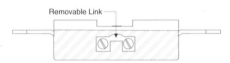

(C) Switched Receptacle

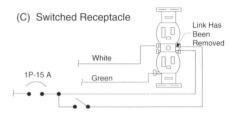

Illustration #109 - Receptacle Types

(B) Split-Bus Receptacle

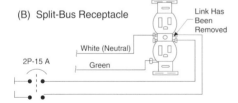

Heavy-Duty Industrial Receptacles

These receptacles are available with ratings up to 400 A, 600 V three-phase but more commonly these receptacles are 30, 60 or 100 A three-phase and either 4 or 5 pole. These receptacles and their mating plugs are designed to meet the rigors and stresses that come from an industrial setting. Both the receptacle and the plug are configured to accept each other if they have the same rating. This is accomplished with a pin and sleeve matching design.

Heavy-Duty Industrial Receptacles (cont'd)

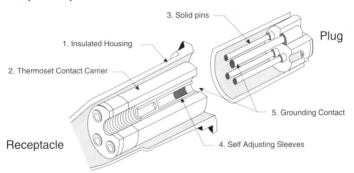

1. Insulated Housing
2. Thermoset Contact Carrier
3. Solid pins
Plug
5. Grounding Contact
4. Self Adjusting Sleeves
Receptacle

1. Insulated Housing
Extremely tough nylon is used for safety and maximum impact strength and abuse resistance.

2. Thermoset Contact Carrier
A thermoset polyester interior is used to resist arcing and prevent damage from overheating.

3. Solid Pins
Solid pin construction provides for long life and reliable electrical contact.

4. Self Adjusting Sleeves
Multi contact inserts ensure easy insertion and constant contact pressure.

5. Grounding Contact
Multi contact insert and solid grounding pin will provide good equipment grounding continuity. Grounding pin is oversized.

Illustration #110 - IEC Rated Pin and Sleeve

Heavy-Duty Industrial Receptacles (cont'd)

IEC 309-1 & 309-2 PIN AND SLEEVE RECEPTACLES

Ampere Rating	2-pole 3-wire			3-pole 4-wire				4-pole 5-wire		
	125 V	250 V	480 V	125 / 250 V	3 Φ 250 V	3 Φ 480 V	3 Φ 600 V	3 Φ Y 120/ 208 V	3 Φ Y 277/ 480 V	3 Φ Y 347/ 600 V
20 A	⊙	⊙	⊙	⊙	⊙	⊙	⊙	⊙	⊙	⊙
30 A										
60 A			NOTE: Receptacles for higher rated currents have similar configurations but are physically larger in size. Plugs are the mirror images of the receptacles. Solid dot represents the ground sleeve position.							
100 A										

Table #44 - Pin and Sleeve Receptacles

Heavy-Duty Industrial Receptacles (cont'd)

Conventional NEMA/EEMAC pin and sleeve designs have served the electrical industry well. However, experience has shown that metal body pin and sleeve products, which have solely relied on a keyway to ensure correct mating, could be incorrectly mated if the keyway were damaged and the devices were forced together. One solution to this problem is to use a ground pin that is larger in dimension than any of the other line-side contact pins (see illustration #110). The receptacle and plug body are made out of nonmetallic materials. Unique pin configurations for each voltage and amperage combination are used. This design is an adopted standard for Europe and is included in the IEC Standard 309-1 and 309-2. Canada and the United States are moving towards adopting this standard, and industrial receptacles based on this standard are now available in North America.

Table #44 lists and shows the North American ratings and configurations for pin and sleeve industrial receptacles and plugs that are in accordance with the IEC 309-1 and 309-2 standard.

Maintenance of Receptacles and Plugs

Receptacles and plugs, as with all electromechanical devices, require periodic inspection and maintenance to assure continued reliability, safety and performance.

Receptacle Maintenance Suggestions:

1. Ensure that a wall plate or box cover is properly installed and secured over the receptacle.
2. Check for cracks, chips or missing parts on the receptacle.
3. Ensure weatherproof cover plates are on receptacles exposed to the weather or in areas where moisture is a problem.

Maintenance of Receptacles (cont'd)

4. Check if face of the receptacle is intact.
5. Check to see that the receptacle is fastened firmly to its box or enclosure.
6. Check that the receptacle being used is the proper one for its intended use and environment and for the duty of operation for which it is rated.
7. Check if the mating pressure of the contacts are adequate.
8. Check the operating temperature of the receptacle by feeling the box cover. Excessively warm temperatures may indicate loose wire terminations or worn contacts.
9. Test GFCI receptacles once a month in accordance with manufacturers' recommendations.
10. If receptacles are in a dusty or dirty environment, occasional cleaning and removal of blanketing elements from the box or enclosure should be done.

11. Check the raised rib and keyway on some industrial receptacles for wear and damage. Replace immediately if damaged or worn.

Plugs and Connectors Maintenance:

1. Check and ensure that the outer jacket of the portable cord or cable is free of cracks or cuts and that the cord insulation is completely inside of the strain relief or cord clamp connector.
2. Check whether the complete cord clamp or strain relief is intact and tightly gripping the cord or cable with all clamping screws, glands, or nuts tightened fully.
3. Check and ensure that the outer diameter of the cord or cable is being used within the range of sizes recommended by the device manufacturer.
4. Check if all the blades, contacts and strain relief of the plug or connector are present, intact and secured firmly.

Maintenance of Receptacles (cont'd)

5. Check if the plug and/or connector housing or enclosure is free of cracks, chips or worn safety pins.

6. Check for the mating pressure of the connector and plug, and replace the devices if adequate pressure is not present.

7. Check the protective covers or lids that may be present on some plugs and connectors. Check gaskets on plugs and connectors that use them.

8. Check the operational duty and application of the plug or connector, and determine whether the device is suitable and has the required rating.

9. Check the ground pin or slot, and check for grounding continuity.

10. Check the raised rib and keyway on some industrial plugs for wear and damage. Replace immediately if wear or damage is present.

Switches

A switch is a device intended for making, breaking or changing connections in an electric circuit under rated load conditions. A switch is not designed to interrupt or break a short circuit. Switches covered in this section will be those limited to general use AC and AC/DC rated at 15 - 30 A. Switches are available in a variety of configurations and ratings. The following ratings and configurations appear on switches:

1. Voltage rating

2. Current rating (maximum continuous current the switch may carry or interrupt)

3. AC or AC/DC (DC rated switches must have very strong contacts and springs together with good insulation due to the nature of DC. When an arc has started between DC switch contacts, the arc will continue to burn and in a short time destroy the switch)

4. Duty (light, heavy, motor)

Switches (cont'd)

5. Horsepower rating (motors have high in-rush currents and therefore the switch must be rated for the additional high currents that will momentarily be present)
6. The number of poles (see illustration #111)
7. The number of closed positions (single-throw, double-throw)
8. Type of approved loads (T rated tungsten filament lamp loads have very high in-rush currents; F rated fluorescent lamp or electric discharge lighting loads are very inductive and produce high voltage transients, motor switch)
9. Test laboratory approval (UL, ULC, CSA)

Types of Switches

NEC article 380, or CEC rules 14-508, 510, 512, and 514 should be consulted on the ratings and requirements of general use switches.

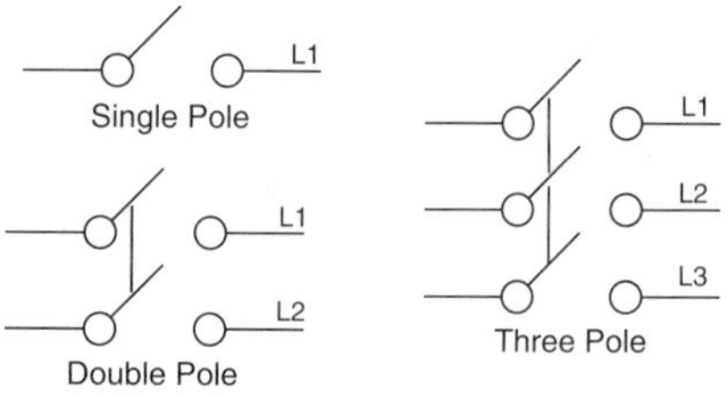

Illustration #111 - Switch Pole Designations

Manufacturers have provided a wide selection of switches that meet the needs of industry. The following illustrations and explanations will provide a view of some of the products that are available:

Toggle or tumbler switch. See illustration #112A.

Push handle switch. Press switch once for on and push again for off. See illustration #112B.

Rocker handle switch. See illustration #112C.

Types of Switches (cont'd)

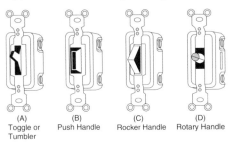

(A)
Toggle or
Tumbler

(B)
Push Handle

(C)
Rocker Handle

(D)
Rotary Handle

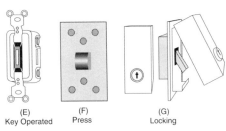

(E)
Key Operated

(F)
Press

(G)
Locking

Illustration #112A-G - Switch Types

Illuminated switch. The handle portion of the switch has a neon lamp.

Rotary handle switch. See illustration #112D.

Key operated. Straight keying. All keys are alike. See illustration #112E.

Press switch. Press switch once for on and push again for off. See illustration #112F.

Locking switch. Cylinder lock with its own unique key. See illustration #112G.

Pendant switch. Used to control lamps or other devices which are mounted overhead and that are connected to a two wire pendant cord. See illustration #112H.

Feed through switch. Designed to be inserted in a portable cord for control of the cord circuit. See illustration #112I.

Surface wiring switch. See illustration #112J.

Types of Switches (cont'd)

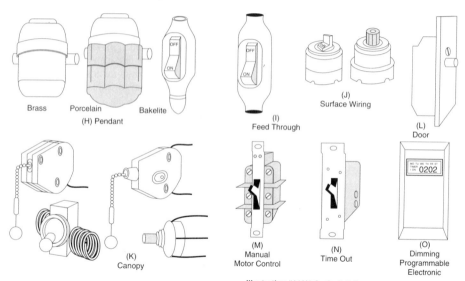

Brass   Porcelain   Bakelite

(H) Pendant

(I) Feed Through

(J) Surface Wiring

(L) Door

(K) Canopy

(M) Manual Motor Control

(N) Time Out

(O) Dimming Programmable Electronic

Illustration #112H-O - Switch Types

Types of Switches (cont'd)

Canopy switch. Small compact switches that are mounted in the canopy of luminaires for the control of lamps. They are available in either pull chain or toggle form. See illustration #112K.

Door switch. See illustration #112L.

Manual motor control switch. Rated in horsepower and may include overload heater element. Available in single, double, or three pole configuration with single or double throw feature. See illustration #112M.

Time-out switch. Switch is manually turned on and automatically turns off in 5, 15 or 30 minutes. Typical applications are closets, stock rooms and other spaces that have a short time occupancy usage. See illustration #112N.

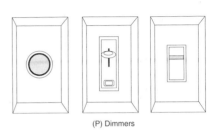

(P) Dimmers

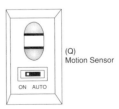

(Q) Motion Sensor

ON   AUTO

Illustration #112P-Q - Switch Types

Types of Switches (cont'd)

Programmable electronic switch. Switch provides multiple on/off cycles and the program period is based on a 7 day week. Switch also has a dimming feature. This switch is limited to noninductive lighting loads. See illustration #112O.

Dimmer switch. Electronic control of source voltage and available for incandescent and inductive lighting loads. Dimmer switches come in a variety of styles including rotary, slider, sensitouch and programmable. See illustration #112P.

Motion sensor switch. Passive infrared sensor detects and responds to changes in radiated heat within the room caused by the presence and movement of a human body. See illustration #112Q.

Electrolier or Trilite switch. Used with dual filament lamps to provide three levels of light for table lamps.

Three way switch. Two position, three terminal switch with a common terminal to provide control from two locations, as shown in illustration #113A.

Four way switch. Two position, four terminal switch designed to provide control from more than two locations and used in conjunction with two three-way switches. See illustration #113B.

Three position momentary contact switch. Double throw momentary contact switch which remains in the off position until pressed in either the top or bottom position. After the chosen position has been released, the switch reverts back to the off position.

Types of Switches (cont'd)
Three position maintained contact
switch. Double throw switch with the top position being on #1, the center position being off, and the bottom position being on #2. Useful when it is desirable to control two loads alternately from one central location: for example a fan and an air conditioner. See illustration #113C.

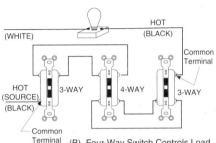

(B) Four-Way Switch Controls Load From More Than Two Locations

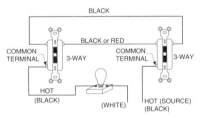

(A) Three Way Switches

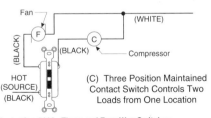

(C) Three Position Maintained Contact Switch Controls Two Loads from One Location

Illustration #113 - Three and Four Way Switches

Switch and Box Dimensions

Switches rated 120/277 V have different physical dimensions than switches rated 347 V, as shown in illustration #114. 120/277 V rated switches must be installed in different boxes than 347 V switches due to the differences in their physical size.

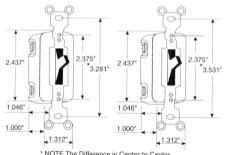

* NOTE The Difference in Center to Center Mounting Strap Length

120/277 V Switch   347 V Switch

Illustration #114 - Switch Dimensions

Switch Maintenance

1. Check whether a wall plate or box cover is over the switch with all attachment screws in place and tightened.
2. Check the condition of the wall plate. Look for cracks, broken parts, etc.
3. Check and make sure the switch is firmly and tightly connected to the box.
4. Check the action of the switch handle. Does it work smoothly and properly or feel loose? Is there any visible arcing?
5. Check the operation of the switch when in the on position. Does the equipment it controls function properly? Is there any flickering of lighting loads, or does a motor start too slow? This may indicate improper operating mechanisms or loose terminal conductor connections.
6. Check the temperature of the switch cover. If it is excessively warm, this could be an indication of loose connections or a faulty operating mechanism.

Switch Maintenance (cont'd)

7. Check motor switches and ensure that the motor properly starts and stops. If this is not occuring the switch contacts may not be fully making and breaking.
8. Check the duty and application of the switch against its present usage and then determine whether the switch has the proper rating.

Note: The frequency of inspections will depend on environmental conditions. It is recommended that inspections be done during regularly scheduled maintenance periods by qualified electrical staff.

Cover Plates

Cover plates are required for receptacles and switches and may be of either the flush or surface type. Cover plates are available in a variety of configurations and in single or multiple gang form, as shown in illustration #115.

Plates are manufactured in different colors and materials, some of which are listed below:

a. Thermoplastic, thermoset (phenolic, urea, etc.) and nylon, in ivory, black, brown, grey, white, blue and red colors.
b. Aluminum in black lacquer.
c. Brass in sandblast, flemish, lemon, oxidized and polished.
d. Bronze in brush, polished and dark.
e. Bakelite in brown, ivory and black.
f. Cadmium in polished form.
g. Copper in brush, mottled, oxidized and polished.
h. Enamel in ivory, brown and black.
i. Nickel in polished form.
j. Silver in brushed, oxidized, polished and satin.
k. Stainless steel in polished and brushed.
l. Wood in various types and stains.
m. PVC in grey.

Cover Plates (cont'd)

Two Gang Combination
Surface Mounted

Three Gang Combination
Flush Mounted

Single Gang
Weatherproof
Duplex Receptacle

Illustration #115 - Cover Plates

Hazardous Locations

Wiring devices located in areas classified as being hazardous must meet the requirements of Articles 500-504 of the NEC or Section 18 of the CEC. Enclosures housing wiring devices must be rated for the classified area in which they are placed. Historically, the NEC and the CEC have set up classes, divisions and groups to help evaluate hazardous locations, to select appropriate equipment and to set out installation standards.

New international manufacturing techniques and standards have been developed that are now being introduced in North America. The IEC international standard defines the guidelines for classifying hazardous locations. Instead of classes, divisions, and groups, the IEC uses the term Zones. In table #45 the North American and international classifications are listed and compared.

Hazardous Locations (cont'd)

Comparison of Hazardous Locations	
North America (NEC/CEC)	**International (IEC)**
*Class 1 Division 1: Location in which flammable gases or vapors are expected to be present during normal operations. *Class 1 Division 2: Location in which flammable liquids or gases are handled, processed or used, but in which they will normally be confined within closed containers or closed systems from which they can only escape in the event of accidental rupture or breakdown. **Class 2 Division 1:** Location in which combustible dust may be in suspension in the air under normal conditions in sufficient quantities to produce explosive or ignitible mixtures. **Class 2 Division 2:** Location in which combustible dust will NOT normally be in suspension, but where accumulation of the dust may interfere with the safe dissipation of heat from electrical equipment or where accumulations near electrical equipment may be ignited by arcs, sparks or heat from the electrical equipment. **Class 3 Division 1:** Location in which readily ignitible fibers or materials producing combustible flyings are handled, manufactured or used. **Class 3 Division 2:** Location in which readily ignitible fibers are stored or handled outside of the manufacturing process.	**Zone 0:** Location in which an explosive gas-air mixture (defined as a mixture of flammable gases or vapors with air, in which, after ignition, combustion spreads throughout the unconsumed mixture) is continuously present or present for long periods. **Zone 1:** Location in which an explosive gas-air mixture is likely to occur in normal operation. The gas-air mixture is considered a probable hazard. **Zone 2:** Location in which an explosive gas-air mixture is NOT likely to occur, and if it does, is only present for a short period of time. **Zone 20:** Location in which dusts or fibers, as a cloud are continuously present during normal operations. **Zone 21:** Location in which dusts on fibers, as a cloud, are likely to occur during normal operation **Zone 22:** Location in which dusts or fibers, as a cloud, can occur infrequently, and persist only for a short period of time.

*Note: Proposed changes in the 18th edition of the CEC adopt the division of Class 1 locations into zones 0, 1, and 2.

Table #45 - North American vs International Classifications of Hazardous Locations

Hazardous Locations (cont'd)

Table #45 lists the types and conditions for various hazardous locations. However there is a further need to evaluate and classify the specific nature of the hazardous substance. This is done by breaking down the gases and vapors of class 1 locations into four groups: Group A, Group B, Group C, and Group D. Groups A-D are grouped according to their ignition temperature, their explosion pressures, their flame propagation nature, and by other flammable characteristics.

The most common gases and vapors encountered in industry are in Group D where gases such as butane, acetone, gasoline, methane, natural gas, and propane are found. Class 2 locations are divided into three groups: Group E, Group F, and Group G. Groups E-G are grouped according to their ignition temperature and/or their electrical conductivity properties.

Group E is comprised mostly of metal dusts, Group F is comprised mostly of coal dusts and Group G is comprised mostly of grain dusts. Class 3 locations have no group classifications.

International/North American Group Comparison		
IEC Groups	**Gases**	**NEC/CEC Groups**
IIA	Propane etc.	D
IIB	Ethylene etc.	C
IIC	Hydrogen	B
	Acetylene	A

Table #46 - Group Comparisons

Hazardous Locations (cont'd)

The IEC (International standard) divides explosive materials into two groups: Group I for all mining applications and Group II for all applications in the surface industry. Group II is further subdivided into three subgroups (A, B and C) and compared to NEC/CEC groups as shown in table #46.

Explosive Ignition Hazards

There are four ways in which electrical equipment may become sources of ignition:

1. By normal operational arcs.
2. By sparks caused by electrical equipment such as motor starters or switches.
3. By high temperatures generated in electrical equipment such as lamps or motors. If the surface temperatures of these electrical devices are above the ignition temperature of the hazardous substance, an explosion or fire could occur.
4. An explosion could be set off by the failure in the operation of a piece of electrical equipment such as the burning out of a lamp socket or the shorting of a terminal.

Class 1 electrical equipment locations must be designed on the assumption that gas or vapor will be present and will gain entry into the electrical equipment. This means there is a possibility that an internal explosion could occur and, therefore, the equipment must withstand the explosion pressure and vent the exploded gases to the surrounding atmosphere at a temperature that is below the ignition point.

See illustration #116 for examples of approved explosion proof equipment.

Note: Some Class I devices come "factory sealed".

Hazardous Locations (cont'd)

Class 2 equipment must be designed to seal out dusts and, therefore, is gasketed. It also must operate below the hazardous substance's ignition temperature and allow for the blanketing by dust of the electrical equipment.

Class 3 equipment must be dust tight and, therefore, is gasketed. It must prevent the escape of sparks from the electrical equipment, and it must operate below the ignition temperature of the hazardous substance.

See illustration #117 for examples of Class 2 and Class 3 equipment.

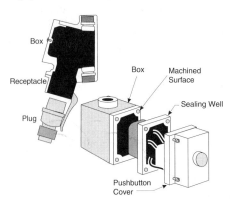

Illustration #116 - Class 1 Devices

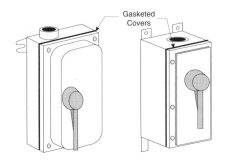

Illustration #117 - Class 2 and 3 Devices

IEC Explosion Proof

Explosion proof equipment designed with the IEC designation of "flameproof" and using components with "increased safety" terminals, which are enclosed in nonmetallic housings, are now available in North America. Flameproof equipment consists of components that are designed to isolate the sources of ignition, such as switches etc., within enclosures that can contain an internal explosion without igniting the surrounding atmosphere. Illustration #118 shows an IEC rated "flameproof" piece of equipment.

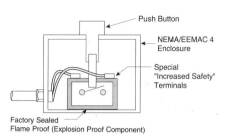

Factory Sealed
Flame Proof (Explosion Proof Component)

"Flameproof" rated equipment are approved in North America for Class 1, Division 1 Groups A, B, C, D, and Class 2, Division 1 Groups E, F, G, and Class 3, Division 1 locations.

Nameplates

All hazardous electrical equipment must carry a proper nameplate clearly showing the class, division, and group as shown in illustration #119.

Illustration #119 - Nameplate for Hazardous Equipment (CSA-UL)

SECTION FIVE

PROTECTIVE DEVICES

Protection Principles

In both the NEC, articles 110 (3, 9 and 10) and 240, and in the CEC, section 14, requirements for electrical circuit protection are mandated. Several factors must be considered when determining the type and amount of protection needed for an electrical circuit.

Perhaps the most important factor or consideration is that electric power systems must be designed to serve loads in a safe and reliable manner. Various forms of protective devices are available to protect electrical circuits, equipment and personnel from injury under abnormal conditions. There are many types of abnormal conditions, some of which are listed and described in table #47.

Abnormal Circuit Conditions	
Type	**Description**
Overload	Currents exceed rated values. Overload values are usually no greater than 6 times rated values.
Short Circuit	Excessive currents at least 6 times above rated values are flowing.
Underload	Current or power flow decreases below a predetermined value.
Undervoltage	Source voltage drops below a tolerable predetermined value.
Overvoltage	Source voltage rises above a tolerable predetermined value.
Reverse Phase Rotation	Source phase sequence is reversed from a predetermined order (eg. ACB instead of ABC).
Phase Unbalance	Polyphase currents are not equal and balanced.
Reverse Current	Currents are flowing out of a circuit instead of into a circuit.
Ground Fault	Electrical circuit is unintentionally in contact with ground or earth.

Table #47 - Abnormal Circuit Conditions

Protection Principles (cont'd)

Section Two of this book (Conductors) stated that when electric current passes through a conductor, heat is generated. The greater the current, the greater the amount of heat generated. The amount of heat generated is a function of the current squared multiplied by time (I^2t). For example, this means that if twice as much current is flowing, then 2^2 or 4 times as much heat is generated; and if 4 times as much current is flowing, then 4^2 or 16 times as much heat is generated.

Section One of this book (Fundamentals) explained that whenever current flows through a conductor, a magnetic field is established around that conductor. The amount of magnetic force present in a current-carrying conductor is a function of the current squared (I^2).

Therefore, if twice as much current is flowing in the conductor, then 4 times as much magnetic force is present, and if 4 times the current is flowing, then 16 times as much magnetic force is present.

Short Circuits and Overloads

Overloads and short circuits impose excessive heating and mechanical stressing on electrical components.

Overheating of conductors creates several problems in an electrical circuit:

1. Conductors which are overheated become annealed or softened. This softening removes the resilience from the conductor and may cause looseness at connection points.
2. Increasing temperatures bring about an increase in the rate of oxidation. Copper oxide is not a good conductor and thus the electrical integrity of a terminal connection is weakened.

Short Circuits and Overloads (cont'd)

2. (cont'd) Aluminum exhibits even higher resistivity when oxidized and thus experiences even greater problems. The oxide is resistive to current flow.

   Therefore both a voltage drop and heating of the terminal will occur. Often enough heat is generated to completely destroy the terminal connection.

3. Overheating acts upon the conductor's insulation and causes it to become dry and brittle. In time the insulation could crack and fall off resulting in exposed conductors. A fire may occur if the exposed conductors touch each other or grounded equipment.

Mechanical stressing of electrical equipment due to short circuits can bring about severe and at times totally destructive results. If any component of the electrical system is not rated for the mechanical forces imposed upon it, the result may be catastrophic.

Conductors, fuses, circuit breakers and distribution equipment have exploded or torn away from their supports by the mechanical forces of high short circuit currents. Fires, explosions, poisonous fumes and panic also accompany uncontrolled short circuits.

To reduce the hazards that accompany a condition of overcurrent, fuses and circuit breakers are installed in the electrical system. These circuit protection devices must be rated for the type and size of load that they are protecting, and must be rated for the fault current capacity of the system in which they are placed. Determining the required ratings of a fuse or circuit breaker may involve complicated and lengthy calculations. Both the NEC and the CEC have published values of minimum and maximum ratings for overcurrent devices. As illustration #120 shows, fuses and circuit breakers must have their ratings clearly marked.

Short Circuits and Overloads (cont'd)

The Code requires that the rated voltage, rated load current and the amperes interrupting capacity (AIC) be shown.

The amperes interrupting capacity (AIC) refers to the highest available symmetrical RMS alternating current which may be safely interrupted by the fuse or circuit breaker at rated voltage without doing damage to itself.

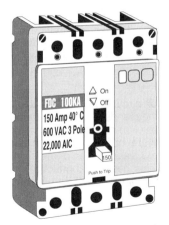

Illustration #120 - Circuit Breaker & Fuse Rating

Short Circuit and Overloads (cont'd)

Electrical distribution systems are often quite large and complicated. Therefore as a result, these systems cannot be made absolutely fail-safe. However, if reliable protective devices are installed, the amount of disruption and damage may be minimized.

Overloads are low level faults that are caused by harmless temporary surge currents that occur when motors are started up, transformers are energized, or they may be continuous overloads due to overloaded motors, transformers, etc.

Note: Despite the magnitude of overloads being between one to six times the normal current level for motors and eight to twelve times for transformers, removal of the overload current within a few seconds will generally prevent equipment damage. Short circuits are much more of a serious problem, because fault currents may be many hundreds of times larger than the normal operating current.

A **short circuit** is current that is out of its normal path. It may be caused by insulation breakdown or due to a faulty connection. During the short circuit, current bypasses the load and the only limiting factor is the impedance of the distribution system upstream from the fault. The impedance of the upstream distribution equipment may be very low, for example in the order of 0.005 Ω. At a voltage of 480 V, the fault current could now be as much as 96,000 A.

If this short circuit is not cut off within a matter of a few thousands of a second, then damage and destruction can become extremely serious. Severe insulation damage, melting of conductors, vaporization of metal, ionization (air particles become conductive) of gases, arcing and fires, and huge magnetic-field stresses that warp or distort electrical equipment all result from large fault currents.

Short Circuit Calculations

Illustration #121 is a representation of an electrical distribution system. If a fault should occur at the point marked "X", how much fault current would flow? To answer this, it must first be determined what the sources of fault current are, what contribution each of these sources makes and what limiting factors to the current flow exist in the fault circuit.

The amount of fault current that will flow to point "X" is dependent upon the following:

1. The size of the transformer: This value is expressed in either kVA or MVA. The larger the transformer, the greater the capacity to feed fault current to the short circuit. A transformer could be likened to a water dam, in that the larger the transformer, or dam, the greater the fault circuit, or volume of water that could flow if the dam broke.

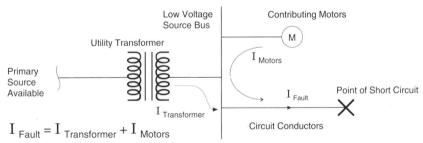

$$I_{Fault} = I_{Transformer} + I_{Motors}$$

Illustration #121 - Fault in Electrical Distribution System

Short Circuit Calculations (cont'd)

2. The impedance of the transformer:
This is the internal opposition that the short circuit current encounters as it passes through the transformer. It is usually expressed in per unit values or as a percent of the rated voltage of the winding (eg. Z = 0.05 or 5%).

Percentage impedance is the percentage of rated primary voltage required to cause full rated load current to flow in the short circuited secondary. Impedances vary with types and designs of transformers, and each transformer will have its own impedance value.

Typical values of %Z are:
500kVA or less = 1.0% to 5%
over 500kVA = 4% to 10%

An example will help to illustrate the meaning and significance of %Z. Illustration #122 shows two transformers with identical kVA ratings but different %Z values. Remembering that %Z refers to the percentage of primary voltage required to produce rated secondary current, it can then be determined what the secondary current would be if rated primary voltage were present by dividing rated secondary current by the decimal equivalent of the %Z.

$I_{secondary} = kVA \times 1000/V_{secondary}$
$I_s = 500 \times 1000/480, \quad I_s = 1042 A$

$I_{secondary}$ is the same for both transformer #1 and transformer #2. However %Z for transformer #1 is 5% and for transformer #2 is 2.5%. Under short circuit conditions the amount of available fault current on the secondary terminals of the transformers is different.

Short Circuit Calculations (cont'd)

Transformer #1 - I_{sc} = 1042/0.05

I_{sc} = 20,840 A

Transformer #2 - I_{sc} = 1042/0.025

I_{sc} = 41,680 A

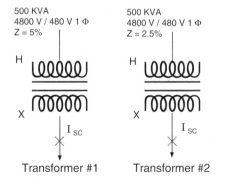

Illustration #122 - Transformers with Different Impedances

Twice as much fault current flows through the transformer with the lower %Z. Therefore, the lower the %Z, the higher the available fault currents.

3. The impedance of the utility source: This value can be expressed in ohms or per unit values based on a common denominator such as the kVA of the transformer, or it can be expressed in available fault capacity such as MVA.

4. The size and type of motor feeding the fault: The larger the motor, the larger the contributing fault current. Induction motors act briefly as generators when de-energized and will in the first couple of cycles of the fault be contributors of fault current. An approximate value of the instantaneous short-circuit current from a motor at an instant 1/2 cycle after the fault has occured is 3.6 times the full load current of the motor. For synchronous motors, the value is 4.8 times the full load current of the motor.

Short Circuit Calculations (cont'd)

5. The impedance of the distribution lines feeding the fault: This value can be expressed in ohms or in per unit values.

6. The time at which the fault occurred in the electrical cycle and the power factor of the fault: Illustration #123 shows two different short circuit waveforms. Illustration #123A is symmetrical; that is the wave has the same magnitude above and below the zero axis. Illustration #123B is asymmetrical; meaning the axis of symmetry is displaced or offset from the zero axis, and the magnitude above and below the zero axis is not the same. The amount of asymmetry during a fault depends on the X/R ratio of the circuit (the poorer the power factor, the greater the asymmetry), and the point on the voltage wave when the fault occurs.

Asymmetrical RMS short circuit current values could theoretically be 1.73 times as large as symmetrical values. However, the usual values are somewhere between 1.25 and 1.5. As can be seen in illustration #123B, this asymmetry is only a factor during the first three to four cycles of the short circuit.

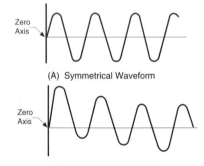

(A) Symmetrical Waveform

(B) Asymmetrical Waveform

Illustration #123 - Short Circuit Waveforms

"Quick" Short Circuit Calculations

Accurately calculating the amount of short circuit current that is flowing into a fault in an electrical system can be very complicated and time consuming. However, there are "Quick Methods" of determining short circuit current values. The following method will result in short circuit values that are higher than will actually occur and therefore this method should be considered a conservative estimation of fault values (note: cable impedances are neglected).

Refer to illustration #124 and follow the steps outlined below:

Step 1: Determine the available fault current coming from the utility. Convert this value into a per unit expression as follows:

per unit Z = kVA of transformer/utility kVA available

per unit Z = 750/250,000

per unit Z = 0.003

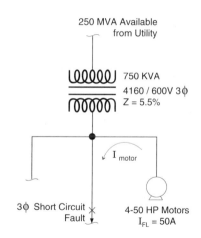

Illustration #124 - Short Circuit "Quick" Method

"Quick" Short Circuit Calculations (cont'd)

Step 2: Convert transformer %Z into per unit values:

per unit Z = %Z/100
per unit Z = 5.5/100
per unit Z = 0.055

Step 3: Add transformer per unit value to utility per unit value.

Total per unit Z = 0.055 + 0.003 ...
$$Z = 0.058$$

Step 4: Calculate symmetrical fault current not including contribution from motors:

$I_{symm.}$= (kVA of Transformer x 1000)
/(V x $\sqrt{3}$ x Z)

$I_{symm.}$ = (750 x 1000)/(600 x $\sqrt{3}$ x 0.058)
$I_{symm.}$ = 12,442 A

Step 5: Add motor contribution:
4 induction motors at 50A full load
4 x 50 x 3.6(motor factor) = 720 A
I_{total} = $I_{symm.}$+ I_{motor}
I_{total} = 12,442 + 720 = 13,162 A

Step 6: Calculate the asymmetrical fault current available (assume a 1.25 multiplier):

$I_{assymmetrical}$ = 1.25 x $I_{symmetrical}$
$I_{asymmetrical}$ = 16,453 A

An even simpler and quicker but still conservative method of calculating short circuit values would be to use the following formula:

1. Single phase system:

$I_{symm..}$ = (kVA of Transformer x
100,000)/(V_{load} x %Z)
eg.$I_{symm.}$= 750 x 100,000/(600 x 5.5)
$I_{symm.}$ = 22,727 A

2. Three phase system:

$I_{symm.}$ = (kVA of Transformer x
100,000)/(V_{load} x $\sqrt{3}$ x %Z)
eg.$I_{symm.}$ = 750 x 100,000/(600 x $\sqrt{3}$ x 5.5)
$I_{symm.}$ = 13,122 A

Note: Using the quick method will usually result in calculated short circuit values being larger than actual values.

Fuses

The fuse is a simple but reliable overcurrent protection device. A low impedance link, or series of links, are encapsulated in an insulating fiber tube or a porcelain cup. The electrical resistance of the link, or links, is very low and acts as a good conductor. However, the links have a lower melting point than the copper or aluminum conductors to which they are connected.

When destructive currents occur, the fuse links respond very quickly and melt. The fuse is designed to be the weak link in the electrical circuit, and should open before any damage occurs to the circuit conductors or to other circuit components and loads. Fuses have some important characteristics that make them an essential and important part of an electrical distribution system:

1. Fuses can have high current-interrupting ratings and therefore can safely withstand very high fault currents without rupturing.

2. Fuses, when correctly sized and coordinated with other circuit protection devices, will prevent needless blackouts. The fuse nearest the fault will open before any upstream devices are affected.

3. Fuses can have the capability to limit the amount of let-through fault current, thereby protecting both upstream and downstream electrical equipment from high fault current stress.

4. Fuses are simple devices that require no periodic servicing or testing.

5. Fuses are standardized according to various UL and CSA classes, thereby leading to greater reliability, consistency, and dependability when specified for a specific application.

6. Fuses are compact and do not have a large space requirement. Purchase costs are low when compared to other circuit protection devices such as circuit breakers and protective relays.

Fuse Ratings

Fuses fall into one of two major groups:
a. low voltage (600 V or less)
b. high voltage (over 600 V)

Within each of these two groups there are many classes of fuses. Fuses are rated according to their ampere rating, voltage rating, and according to their current interrupting rating.

Current Rating: Every fuse has a specific ampere rating. In selecting the ampere rating of the fuse, consideration must be given to the type of load being protected, the fuse size allowed by Code (NEC/CEC), and the degree of selective coordination required.

Voltage Rating: The voltage rating of the fuse must be at least equal to and preferably greater than the circuit voltage. Fuses that are rated higher than the circuit voltage may be used for lower voltage applications.

The voltage rating reflects the ability of a fuse to suppress the internal arcing that occurs after a fuse link melts and an arc is produced. *If the fuse does not have at least the voltage rating of the circuit and an overcurrent condition should occur, the ability of the fuse to suppress the arc cannot be guaranteed. This is clearly an unsafe situation and must be avoided at all times.*

Current Interrupting Rating: A fuse must be able to withstand the destructive energy that accompanies an overcurrent condition. If the fault current is greater than the fuse is rated for, then the fuse could either rupture, explode or discharge destructive flames and gases.

Fuse Ratings (cont'd)
Current Interrupting Rating (cont'd):

It is a Code requirement that the overcurrent device be capable of withstanding and interrupting the maximum amount of fault current that is available in the system. The rating which defines the capacity of a protective device to maintain its electrical and mechanical integrity when reacting to fault currents is known as its "ampere interrupting rating" (AIC).

Current Limiting

There are two terms that are frequently used when discussing current limitation. The first term, I^2t refers to the thermal energy of a fault and the second term, I^2 refers to the electromagnetic stresses present during a fault. The amount of I^2t and I^2 available under a large short circuit can be extremely high, as seen in illustration #125A.

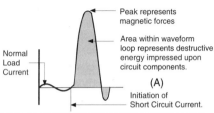

A non-current limiting protective device by permitting a short circuit current to build up to its full value could let a large amount of destructive short circuit heat energy through before opening the circuit.

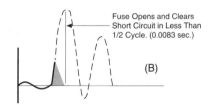

A current limiting fuse has a high speed of response and cuts off a short circuit current long before it can build up to its full peak value.

Illustration #125 - Thermal & Magnetic Energy

Current Limiting (cont'd)

However, if a fuse can interrupt the fault current within an extremely short period of time (0.004 of a second), then the extent of thermal and magnetic energy will be greatly curtailed as seen in illustration #125B.

If a protection device interrupts a fault current in less than one half cycle and before it reaches its total available peak value, it is then classified as a current limiting device. Most fuses are current limiting, and that is one reason why fuses are commonly used in electrical systems where large fault currents may be present. Current limiting fuses restrict faults to such low values that a high degree of protection is given to circuit components. These fuses allow for the selection of circuit equipment with much lower interrupting levels to be used in the electrical system, even though high levels of fault current are available.

Log-Log Graphs

Fuse manufacturers publish information on each of their current limiting fuses and provide graphs as shown in illustrations #126 and #127. The log-log graph depicts current limiting values corresponding to each fuse.

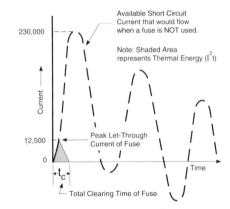

Illustration #126 - Action of a Current Limiting Fuse

Current Limiting/Log-Log Graph
(cont'd)

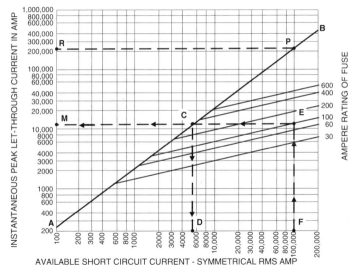

Illustration #127 - Current Limiting Characteristics

Current Limiting/Log-Log Graph (cont'd)

To interpret the log-log graph of illustration #127, key areas of the graph need to be identified and explained:

The horizontal axis: This is labelled the "available short circuit current - symmetrical RMS amp", and serves as the starting point to enter the graph.

For example, assume the electrical system has the capacity to provide 100,000 A to an electrical fault and that a 100 A current limiting fuse is being used. First locate 100,000 A on the horizontal axis (shown in illustration #127 as *point F*). The dotted vertical line intersects the 100 A fuse line *(point E)* at a vertical axis value of 12,500 A *(point M)*. This 12,500 A value represents the *PEAK* instantaneous current that will occur, and is reflective of the magnetic forces that will be imposed upon the electrical circuit (I^2).

The dotted horizontal line *EM* also intersects the line *AB* at point *C*. The line drawn vertically to the horizontal axis from point *C*, and intersecting at point *D*, represents the value of apparent RMS symmetrical let-through current by the fuse during its current limiting operation. In the case of illustration #127, this value is 5,200 A. The I^2t of 5,200 A is representative of the thermal energy released into the protected circuit.

The vertical axis: This is labelled "instantaneous peak let-through current in amps". This axis provides a value of the worst case peak current that would be reached if a certain amount of RMS short circuit current were available. From illustration #127, if 100,000 A of RMS fault current were available, the maximum *PEAK* current that could be present on the first cycle of the fault is found at *point P* (intersection of line *AB* and the 100,000 A vertical line starting at point *F*).

Current Limiting/Log-Log Graph (cont'd)

From illustration #127, this value is 230,000 A (**point R**) and is also shown in illustration #126. This magnitude represents the asymmetrical peak current based on a fault power factor of 0.15 and with the fault occurring when the AC voltage waveform is passing through zero.

Peak asymmetrical values are derived from two factors:

1. The time at which the fault occurred.
2. The power factor of the fault (the theoretical maximum **peak** amperes at half cycle is 2.828 times the RMS symmetrical value).

Selective Coordination of Fuses

Careful analysis of the fault clearing times of each overcurrent device and proper selection of fuses and circuit breakers will prevent unnecessary power outages.

When the protective device nearest a faulted circuit opens, as seen in illustration #128, and larger upstream fuses or circuit breakers remain closed, then the electrical system is selectively coordinated.

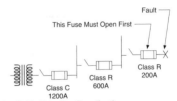

Illustration #128 - Selective Coordination

Fuses are selectively coordinated when the total clearing energy on the load side fuse is less than the melting energy on the line side fuse. Fuse manufacturers publish recommended ratios between upstream and downstream fuses. **Each manufacturer has printed information (see example in illustration #129) on each of its fuses which should be consulted for proper selective coordination.**

Selective Coordination of Fuses (cont'd)

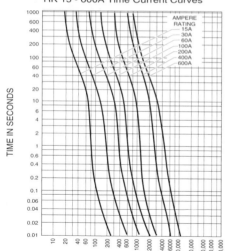

RK 15 - 600A Time Current Curves

AMPERE RATING
15A
30A
60A
100A
200A
400A
600A

TIME IN SECONDS

CURRENT IN AMPERES

Illustration #129 - Time Current Fuse Curves

Low Voltage Fuse Classes

UL and CSA have developed basic performance and physical specifications or standards for fuses which has resulted in the establishment of distinct classes of low voltage fuses. UL and CSA standards are very similar but do differ for some fuse classes.

The following fuse guide, table #48, includes a comprehensive listing of class descriptions, ratings and illustrations.

Low Voltage Fuse Classes (cont'd)

Low Voltage Fuses					
Type	Standards	Voltage Rating	Current Rating	Interrupting Rating (AIC)	Comment
Micro	UL 198G CSA C22.2 No.59.2	125 V, AC	up to 10 A	50 A symmetrical	
Miniature		125/250 V, AC	up to 30 A	10,000 A symmetrical	
Misc. Cartridge		125 - 600 V, AC	up to 30 A	10,000, 50,000 or 100,000 A symmetrical	

Micro Fuse

Cartridge Type Fuses

BLS 5

Table #48 - Low Voltage Fuses

Low Voltage Fuse Classes (cont'd)

Low Voltage Fuses					
Type	Standards	Voltage Rating	Current Rating	Interrupting Rating (AIC)	Comment
Special Purpose	Approved by Application	up to 1000 V	up to 6000 A	200,000 A symmetrical	
	Cable Limiter Welder Limiter Semiconductor Fuse				
Plug	UL 198F CSA C22.2 No.59.1	125 V, AC	up to 30 A	10,000 A symmetrical	Non-renewable, non-current limiting
	Edison Type S Type S to Edison Adapter				

Table #48 (cont'd) - Low Voltage Fuses

Low Voltage Fuse Classes (cont'd)

Low Voltage Fuses					
Type	Standards	Voltage Rating	Current Rating	Interrupting Rating (AIC)	Comment
Class CC (HRCI-MISC)	UL 198C CSA C22.2 No.106	600 V, AC	up to 30 A	200,000 A symmetrical	Non-renewable, current limiting
	$\frac{13}{32}$ in. $1\frac{1}{2}$ in. $\frac{1}{2}$ in.				
Class G	UL 198C CSA C22.2 No.106	300 V, AC to ground	up to 60 A	100,000 A symmetrical	Non-renewable, current limiting
	Class G $\frac{13}{32}$ in. L		up to 15 A L = 1 $\frac{5}{16}$ in. 16 - 20 A L = 1 $\frac{13}{32}$ in. 21 - 30 A L = 1 $\frac{5}{8}$ in. 31 - 60 A L = 2 $\frac{1}{4}$ in.		

Table #48 (cont'd) - Low Voltage Fuses

Low Voltage Fuse Classes (cont'd)

Low Voltage Fuses					
Type	Standards	Voltage Rating	Current Rating	Interrupting Rating (AIC)	Comment
Class H	UL 198B CSA C22.2 No.59.1	250 & 600 V, AC	up to 600 A	10,000 A symmetrical	Renewable & non-renewable, non-current limiting
	up to 60 A — H — Ferrule Type / Knife-Blade — H — 225 - 600 A ONLY / 70 - 200 A 225 - 600 A				
Class J	UL 198C CSA C22.2 No.106	600 V, AC	up to 600 A	200,000 A symmetrical	Non-renewable, current limiting
	up to 60 A — J — Ferrule Type / 70 - 600 A — J — Knife-Blade				

Table #48 (cont'd) - Low Voltage Fuses

Low Voltage Fuse Classes (cont'd)

Low Voltage Fuses					
Type	Standards	Voltage Rating	Current Rating	Interrupting Rating (AIC)	Comment
Class K	UL 198D No CSA Standard	250 & 600 V, AC	up to 600 A	50,000 100,000 200,000 A	Non-renewable, current limiting
	up to 60 A — K — Ferrule Type		70 - 600 A — K — Knife-Blade	Marked either Class K1, K5 or K9 depending on I_p and I^2t values	
Class L	UL 198C CSA C22.2 No.106	600 V, AC	601 - 6000 A	200,000 A symmetrical	Non-renewable, current limiting
	— L —			Mounting holes vary according to fuse ampere classifications	

Table #48 (cont'd) - Low Voltage Fuses

Low Voltage Fuse Classes (cont'd)

Low Voltage Fuses					
Type	Standards	Voltage Rating	Current Rating	Interrupting Rating (AIC)	Comment
Class R	UL 198E CSA C22.2 No.106	250 & 600 V, AC	up to 600 A	200,000 A symmetrical	Non-renewable, current limiting: RK1-high, RK5-moderate
	up to 60 A Ferrule Type		70 - 600 A Knife-Blade		
Class T	UL 198H CSA C22.2 No.106	300 V, AC to ground 600 V, AC	up to 1200 A at 300 V, up to 800 A at 600 V	200,000 A symmetrical	Non-renewable, current limiting
	300 V, up to 60 A 600 V, up to 30 A Ferrule Types		600 V, 35 - 60 A	300 V, 70 - 1200 A 600 V, 70 - 800 A	

Table #48 (cont'd) - Low Voltage Fuses

Low Voltage Fuse Construction

Basic Single Element Fuse: Illustration #130 describes the construction and operation of a basic single element fuse.

Sustained **overload** melts a section of the fuse link, establishes an arc and opens the fuse

Cutaway view

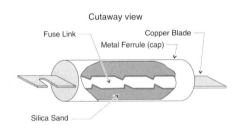

Fuse Link
Copper Blade
Metal Ferrule (cap)
Silica Sand

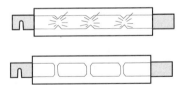

When subjected to a **short circuit** current several sections of the fuse melt almost instantly and open the fuse

Illustration #130 - Basic Single Element Fuse

Low Voltage Fuse Construction (cont'd)

Dual Element Fuse: Illustration #131 depicts and explains the basic construction and operation of a dual element fuse.

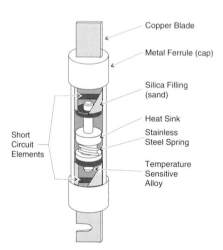

Copper Blade

Metal Ferrule (cap)

Silica Filling (sand)

Heat Sink

Stainless Steel Spring

Temperature Sensitive Alloy

Short Circuit Elements

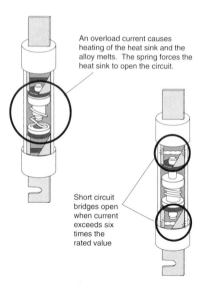

An overload current causes heating of the heat sink and the alloy melts. The spring forces the heat sink to open the circuit.

Short circuit bridges open when current exceeds six times the rated value

Illustration #131 - Dual Element Fuse

Fuses Rated Over 600 V

Power fuses rated over 600 V are either of the current limiting type or of the expulsion type (see illustrations #132 and #133). The current limiting type are categorized as being either general purpose fuses (E rated or non-E rated) or R rated. The E rated fuse is mainly used for transformer primary protection, protection of potential transformers and feeder circuit protection.

Fuses rated 100E or less open within five minutes when loaded to 200-240% of the E (ampere) rating. Fuses rated over 100E open within ten minutes at 220-264% of the E (ampere) rating. Ratings range from 1/4E to 750E, and in voltage ratings of 2750, 5500, 8300 and 15,500 V.

The type E fuse is not intended to provide protection against overload currents because it can only reliably interrupt currents twice above its continuous E rating and for non-E rated fuses, at currents three times above the continuous fuse rating.

R rated fuses are current limiting fuses used specifically for short circuit protection of medium voltage motors (over 600 V).

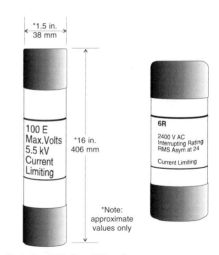

Illustration #132 - E and R Type Fuses

Fuses Over 600V (cont'd)

Multiplying the R rating by 100 yields the ampere level at which the fuse will melt in about 25 seconds.

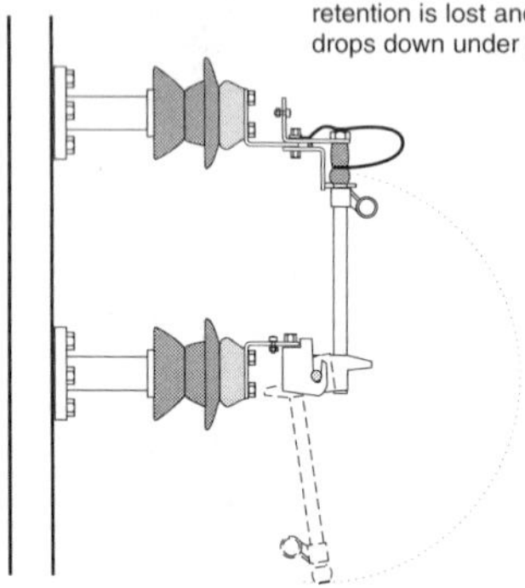

R rated fuses are available in voltages of 2400, 5000, and 8300 V and in current ratings of 2R to 24R. This fuse provides short circuit protection but must be used in combination with an overload protective device to sense and clear overloads.

Expulsion types of fuses are generally used in distribution system cutouts or in disconnect switches. An arc confining tube in combination with a deionizing fiber liner and fusible element is used to interrupt a fault current (see illustration #133). Arc interruption is accomplished by the rapid production of pressurized gases within the fuse tube which then extinguishes the arc by expulsion from the open end(s) of the fuse. The operation of the expulsion fuse is noisy due to the rapid release of gases; proper venting should be in place.

The fuse is spring-loaded into a fuseholder. When the fuse opens due to an electrical fault, spring retention is lost and the fuse drops down under gravity.

Illustration #133 - Expulsion Type Fuses

Circuit Breakers

A circuit breaker is a device that is designed to manually open and close a circuit, and to open automatically on a pre-determined overcurrent, without damage to itself, when properly applied within its ratings.

Low voltage (600 V and less) breakers are divided into two groups: Low voltage power circuit breakers (LVPCB), and molded case circuit breakers (MCCB). Medium voltage breakers (over 600 V and up to 38 kV) are divided into four groups: Air, vacuum, SF6 and oil. All circuit breakers are rated according to their voltage, continuous current, current interrupting capacity (AIC) frequency and operating requirements (temperature, moisture, altitude, mounting position, mechanical shock, reverse feed, etc.). Terminology and principles used when discussing fuses such as selectivity, current limiting, interrupting capacity and fault energy (I^2t, I^2) also apply to circuit breakers.

Molded Case Circuit Breaker (MCCB) Low Voltage

The most commonly used low voltage circuit breaker is the molded case circuit breaker. Illustration #134 provides a cutaway view of the basic parts of a molded case circuit breaker. Molded case circuit breakers are available in voltage ratings of up to 600 V, current ratings from 15 to 3000 A and current interrupting ratings from 10,000 to 200,000 A.

The purpose of the molded case is to provide an insulated enclosure on which to mount all of the circuit breaker components. Each type of molded enclosure is assigned a frame designation in order to provide identification. This frame identification refers to some of the characteristics of the breaker such as voltage, current, interrupting capacity and physical dimensions. In contrast to the standardization of fuse classes, there is no standardization of frame sizes by circuit breaker manufacturers.

Molded Case Circuit Breaker (cont'd)

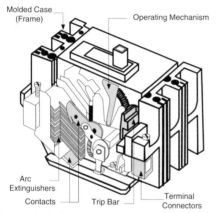

Illustration #134 - Molded Case Circuit Breaker

The operating mechanism shown in illustration #134 provides a means of opening and closing the breaker.

This quick make, quick break mechanism opens and closes at a speed that is independent of the speed at which the operator initiates it.

The molded case circuit breaker is also *trip free* which means that the breaker cannot be prevented from tripping by holding the breaker handle in the "ON" position. When the breaker does trip, the handle moves to the center position requiring the operator to first move the handle to the "OFF" position, and then to the "ON" position in order to reset the breaker. The purpose of the arc extinguishers is to confine, divide and extinguish the fault or load arc drawn between the breaker contacts each time a breaker interrupts current. Standard molded case breakers use an electro-mechanical means to interrupt overloads and short circuits. As shown in illustration #135, a bi-metallic element is used to provide overload protection. In illustration #136 an electro-magnet is employed to provide short circuit protection.

Molded Case Circuit Breaker (cont'd)

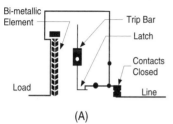
(A)

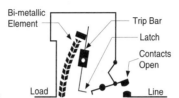
(B) Bi-metallic Element Heats and Bends to Open Contacts on Overload

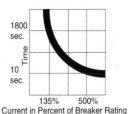
(C) Typical Trip curve for Thermal Action

Current in Percent of Breaker Rating

Illustration #135 - Thermal Bi-Metallic Element (Overload)

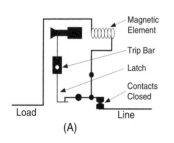
(A)

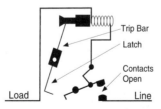
(B) Magnetic Element Closes Gap and Opens Contacts on Short Circuit

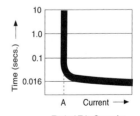
(C) Typical Trip Curve for Fixed Magnetic Action

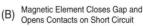
Illustration #136 - Magnetic Action (Short Circuit)

Molded Case Circuit Breaker (cont'd)

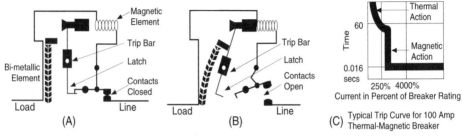

(A)   (B)   (C) Typical Trip Curve for 100 Amp Thermal-Magnetic Breaker

Illustration #137 - Thermal-Magnetic Action

Illustration #135C is a graph based on the inverse time (device operates faster as current increases) characteristics of a thermal overload element. Bi-metallic elements provide a long time delay on light overloads but yet have a fast response on heavier overloads.

Illustration #136C is a graph of a fixed time trip curve. There is no intentional delay in the tripping of the breaker after a certain short circuit current value has been reached.

Illustration #137 combines the two elements and forms a typical thermal-magnetic breaker. Notice the trip curve (illustration #137C) now reflects the influence of each element on the shape of the trip curve.

Molded case circuit breakers are manufactured and available in a variety of types: thermal-magnetic, magnetic, integrally fused, current limiting, solid state and ground fault circuit interrupters.

Molded Case Circuit Breaker (cont'd)

Magnetic circuit breakers (illustration #138) provide short circuit protection only and are equipped with adjustable trip settings. They are commonly used to provide short circuit protection for motors and welders.

Integrally fused circuit breakers are used in systems where very high short circuit currents exist, and the rating of the circuit breaker is not sufficient to handle these currents. The fuses used will limit the short circuit currents to values tolerated by the circuit breaker. This circuit breaker is of the thermal-magnetic type. See Illustration #139.

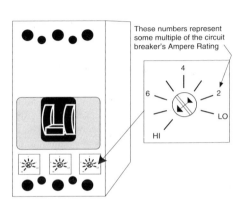

Illustration #138 - Magnetic Circuit Breaker

Molded Case Circuit Breaker (cont'd)

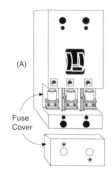

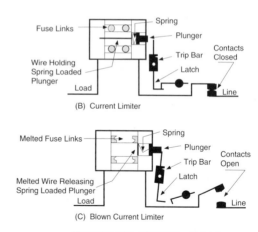

(B) Current Limiter

(C) Blown Current Limiter

Illustration #139 - Integrally Fused Breaker

Molded Case Circuit Breaker (cont'd)

Current limiting circuit breakers are designed to respond to the electromagnetic repulsion created by closely spaced, parallel contact arms carrying current in the opposite direction.

This repulsion causes the contacts to blow open in less than 0.001 of a second when under very high fault current levels. This high speed action limits the amount of I^2t and I^2 energy and force (see illustration #140).

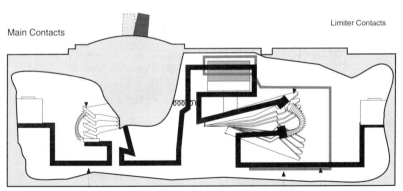

Main Contacts

Limiter Contacts

Conventional Current Path

Limiter Resistor

Limiter Current Path

Illustration #140 - Current Limiting Breaker

Molded Case Circuit Breaker (cont'd)

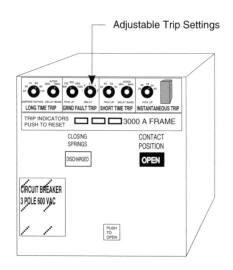

Adjustable Trip Settings

Illustration #141 - Solid State Trip Breaker

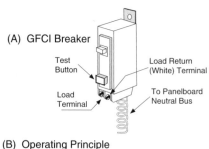

(A) GFCI Breaker

Test Button

Load Terminal

Load Return (White) Terminal

To Panelboard Neutral Bus

(B) Operating Principle

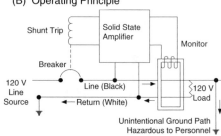

Shunt Trip

Solid State Amplifier

Monitor

Breaker

120 V Line Source

Line (Black)

Return (White)

120 V Load

Unintentional Ground Path Hazardous to Personnel

Illustration #142 - Ground Fault Breaker

Molded Case Circuit Breaker (cont'd)

Solid state trip circuit breakers have the flexibility and versatility to have adjustments made to the following: Continuous current rating, instantaneous pickup rating, ground fault current rating and short time pickup rating. These breakers are available only in the larger frame sizes (800 - 3000 A). See Illustration #141.

Ground fault circuit interrupters are thermal magnetic breakers which incorporate a solid state ground fault sensing circuit to detect ground currents. GFCI's monitor the balance between the current in the line or ungrounded conductor and the current in the return or grounded conductor. If the imbalance exceeds 5 mA, the sensing portion of the circuit breaker will cause immediate interruption of the circuit. GFCI breakers are available in single and two pole design with 15 to 60 A ratings and also provide overload and short circuit protection.

GFCI applications include protection of human life near swimming pools, bathrooms, outdoor receptacles and marinas (see illustration #142).

Molded Case Circuit Breaker Accessories

Numerous accessories are available for molded case circuit breakers:

1. Shunt trip: Used to electrically trip a circuit breaker from a remote location. Consists of a momentary rated solenoid tripping device mounted in the breaker and energized by either an AC or DC voltage.

2. Undervoltage release: An electro-mechanical device which consists of a spring and solenoid mounted on the inside of the circuit breaker. When the voltage falls below 40 to 60% of the solenoid coil rating, the solenoid will cause the breaker to trip. The breaker cannot be turned on until the voltage has returned to at least 80% of its rated value.

Molded Case Circuit Breaker Accessories (cont'd)

3. Auxiliary contacts: Normally open and/or normally closed contacts mounted in the breaker that will change position whenever the breaker is opened or closed.

4. Alarm contacts: Normally open contacts that close only when the breaker has tripped due to an overcurrent condition.

5. Motor operators: Electrically driven motor unit that has an operating arm attached to the breaker handle allowing for remote control of the breaker.

6. Mechanical interlocks of which several types are available: Walking beam type, sliding bar type and key interlock type. The purpose of an interlock is to prevent the simultaneous operation of two breakers. One of the breakers may be closed but the other must remain open.

Low Voltage Power Circuit Breaker (600 V or less)

Power circuit breakers are of open construction metal frame assembly, with all parts designed for accessible maintenance, repair, and ease of replacement.

These breakers are intended for service in switchgear compartments, or other enclosures of dead front construction, as compared to molded case circuit breakers which are intended to be installed in panelboards.

Tripping units, either of the electromagnetic overcurrent direct-acting type or the solid state are field adjustable over a wide range and are interchangeable within their frame size.

Power Circuit Breakers (cont'd)

Power circuit breakers of the drawout (withdrawable) construction may be integrally connected to current limiting fuses in order to obtain high interrupting values of 200,000 A. The operation of these breakers under short circuit conditions involves magnetic repulsion of contacts and an accompanying blast of air that moves the arc into the ceramic arc chutes and then extinguishes it.

Power breakers are available in the following ratings: *Voltage ratings* - 240, 480, 600 V AC; or 240 V DC; *Poles* - 2, 3; *Continuous current rating* - 225 to 6000 A; *Interrupting current rating (AIC)* - 14,000 to 130,000 A unfused, and 200,000 A fused.

Power breakers are used to protect low voltage distribution systems, motor control centers, large transformers, and low voltage alternators. They are designed to allow for coordination and selectivity within a distribution network.

Medium Voltage Circuit Breakers (over 600 V - 38 kV)

The American National Standards Institute (ANSI) requires that all medium voltage circuit breakers be rated and approved for their voltage class and application.

Medium voltage circuit breakers are rated in the following way:

1. Rated maximum voltage: The maximum nominal system voltage expressed in kV RMS at which the breaker operates.
2. Rated continuous current: The maximum load current at 60 Hz expressed in RMS amperes that the breaker may normally carry without overheating.
3. Rated short circuit current: The highest current at rated voltage which the breaker is guaranteed to interrupt without damaging itself.

Medium Voltage Circuit Breakers (cont'd)

4. Rated interrupting time: Expressed in electrical cycles, and is the time that it will take the breaker to open the circuit and interrupt the fault current.

5. Rated short time current carrying capability: The maximum current which the breaker may carry for a specified short time interval, usually three seconds.

6. Rated momentary current: Refers to the maximum amount of available fault current in the system to which the breaker may be subjected without suffering harm due to destructive mechanical stresses (I^2). The momentary current rating reflects the physical or mechanical capability of the breaker.

There are four common types of medium voltage circuit breakers: Air, vacuum, SF_6 and oil. All of these use the two fundamental design principles:

1. Contacts must open as quickly as possible, usually 5 cycles or less.

2. A means of cooling the arc and increasing the insulation between the contacts should be introduced.

Air Circuit Breakers: Whenever two mated contacts carrying current separate, an arc is formed. If the contacts are moved apart slowly, the arc will ionize the air between them. The contacts must then be moved a considerable distance before the arc is extinguished. The distance in contact parting is reduced if the ionized air is removed by blowing a stream of air between the contacts.

Medium Voltage Circuit Breakers (cont'd)

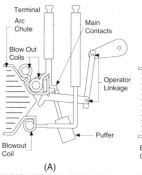

(A)

CLOSED POSITION of Breaker With Current Flowing Through the Terminals and Main Contacts.

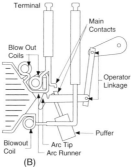

(B)

STARTING TO OPEN, Main Contacts Part and the Current Shifts to Arcing Contacts and Blowout Coils.

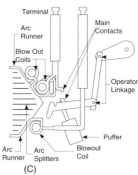

(C)

INTERRUPTION Occurs as Arc is Driven against Arc Splitters by Magnetic Action of Blowout Coils.

Illustration #143 - Medium Voltage Air Circuit Breaker

Medium Voltage Circuit Breakers (cont'd)

Special designs and modifications are employed in the contact breaking area of the circuit breaker. A blowout coil connected in series with the arc is indicated in illustration #143. Arc chutes/splitters are shown in illustration #144. Compressed air blasted across the arc on opening is shown in illustration #145. All of these methods are means by which current may be interrupted.

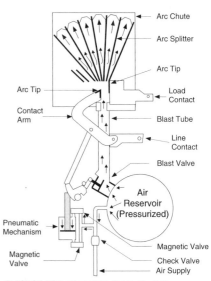

BLAST OF AIR is released simultaneously when contacts are opened and extinguishes the arc rapidly.

Illustration #145 - Compressed Air

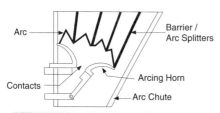

INTERRUPTION of Arc in Typical Air Breaker is Accomplished When Moving Arc Strikes Barriers. Arc Spreads Out to Horn Tips.

Illustration #144 - Arc Chutes/Splitters

Medium Voltage Circuit Breakers (cont'd)

Air circuit breakers have several notable advantages when compared to oil circuit breakers:

1. The risk of fire and the cost of maintenance associated with oil circuit breakers is eliminated.

2. Arc energy products are blasted to atmosphere, and on subsequent interruptions new air is used. This is in contrast to an oil breaker where the same oil is reused in its partly deteriorated condition.

3. Arc energy air pressure is vented outside of the breaker. In oil breakers heavy internal mechanical stresses are set up by gas pressure and oil movement during an interruption.

4. There is an economic advantage over oil circuit breakers. Costs are reduced by eliminating the use of large quantities of expensive oil and the necessity of oil containment.

Vacuum Circuit Breakers: Advances in vacuum breaker technology have resulted in wide acceptance of their use in industry. The advantages offered by vacuum switching makes it an economically efficient and safe means to interrupt medium voltage currents. The basic design, as seen in illustration #146, consists of an arcing chamber with two stem-connected contact pieces located between two ceramic insulators. One contact piece is fixed to the housing while the other is a moving contact piece connected to the housing by vacuum-tight bellows.

The arcing chamber acts as a vapour shield. On opening, a metal vapour arc is drawn between the contact pieces and is then extinguished at current zero. The small amount of metal vapour that is not redistributed over the contact pieces condenses on the arcing chamber wall; this maintains the insulation integrity of the ceramic insulators. The metal bellows allow the moving contact piece to carry out its stroke.

Medium Voltage Circuit Breakers (cont'd)

The moveable contact piece has only to move a short distance (approximately 16 mm at 24 kV). This helps to reduce the overall space occupied by the breaker.

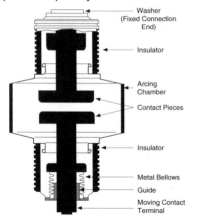

Illustration #146 - Vacuum Circuit Breaker Bottle

Advantages of vacuum circuit breakers:

1. Rated life is long: 10,000 make-break operations at full load current without failure.
2. The arcing chamber is completely enclosed and sealed.
3. Very short arcing and total break time; three cycles.
4. No fire risk.
5. No noise and no emission of harmful gas or air during operation.
6. Vacuum bottles are replaceable, allowing for ease and speed in maintenance.

SF_6 Circuit Breakers: SF_6 (sulfur hexafluoride) has excellent arc quenching and insulating properties. Compressed SF_6 has a very high insulation rating and is used in circuit breakers from 6.6 kV to 765 kV. Its superior arc quenching ability comes from the fact that it is electronegative, which means that it traps electrons from the arc, forming negative ions which are ineffective as current carriers.

Medium Voltage Circuit Breakers (cont'd)

The SF_6 basic design, as seen in illustration #147, consists of a pressure tight casing with SF_6 gas forced into the arc by the motion of a piston attached to the main contact system. The SF_6 gas in the contact chamber is sealed at approximately 35 psig.

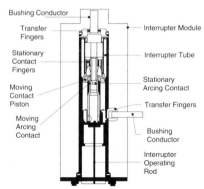

Bushing Conductor
Transfer Fingers
Interrupter Module
Stationary Contact Fingers
Interrupter Tube
Stationary Arcing Contact
Moving Contact Piston
Transfer Fingers
Moving Arcing Contact
Bushing Conductor
Interrupter Operating Rod

Illustration #147 - SF$_6$ Circuit Breaker

Advantages of SF$_6$ circuit breakers:

1. The low gas velocity and the pressures used minimize any tendency towards current chopping. Capacitive currents may be interrupted without restriking.
2. The closed circuit gas cycle and the low gas velocity provide a quiet operation with no exhaust to the atmosphere.
3. The closed gas circuit keeps the interior dry so there are no moisture problems.
4. The excellent arc extinguishing properties of SF_6 gas result in very short arcing times, therefore contact erosion is minimized. Contacts may also be operated at higher temperatures without deteriorating.
5. No carbon deposits are formed. Therefore, tracking or insulation breakdown is eliminated.
6. Electrical clearances are smaller due to the insulating property of the gas.
7. The breaker is totally sealed and enclosed.

Medium Voltage Circuit Breakers (cont'd)

Oil Circuit Breakers: Two types of oil breakers have been manufactured: The bulk oil type which is no longer made, and the minimum oil type which is still available. The oil in the breaker is used to act as an arc extinguishing agent during interruption, to insulate live parts from grounded portions of the breaker and to insulate the contacts from each other. The oil used in circuit breakers has a complex molecular structure consisting mainly of hydrogen and carbon atoms.

Illustration #148 indicates that when the contacts of an oil circuit breaker part, the heat of the arc decomposes the oil and a gas bubble composed mainly of hydrogen forms around the arc. Hydrogen acts as both an arc extinguishing agent and as a means to cool the arc. Minimum oil circuit breakers, as shown in illustration #149, are used mainly for high voltage utility applications and are manufactured for voltages from 2.4 to 240 kV.

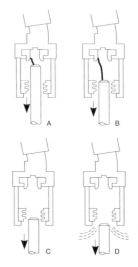

A. Bayonet contact moves down, draws an arc and causes a high pressure gas bubble to form.

B. As the bayonet reaches the helical throat, oil surges in and quenches the arc.

C. High pressure gas causes quick recovery as the gas bubble escapes.

D. Oil enters to refill the chamber.

Illustration #148 - Oil Circuit Breaker

Medium Voltage Circuit Breakers (cont'd)

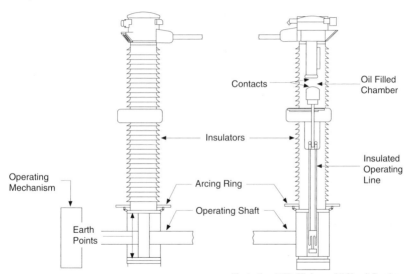

Contacts

Oil Filled Chamber

Insulators

Insulated Operating Line

Operating Mechanism

Arcing Ring

Operating Shaft

Earth Points

Illustration #149 - Minimum Oil Circuit Breaker

SECTION
SIX
DISTRIBUTION

Distribution Equipment Clearances

Electrical distribution equipment encompasses a large and varied number of components. Included in this list are:

1. Metering equipment: meters, meter centers, meter transformers
2. Panelboards: lighting and power
3. Motor control centers: low voltage and medium voltage
4. Switchgear: metal enclosed, metal clad, arc proof, indoor low voltage, indoor and outdoor medium voltage

NEC articles 110-16 (600 V or less) and 110-34 (over 600 V), and CEC rule 2-308, state the need and define the values of clearances required around electrical distribution equipment. Equipment clearances and their conditions are listed in table #49 for NEC, and table #50 for CEC.

In addition to working space in front of or behind electrical distribution equipment, proper ceiling clearances and accessibility to and from electrical equipment rooms or areas are required. Electrical distribution equipment should be located in dry and ambient temperature neutral rooms. Adequate ventilation including filtering of dusts and plant contaminants and humidity control are important.

Each piece of electrical distribution equipment during its normal operation produces heat. If the electrical room is not properly ventilated, elevated temperatures may be encountered which will reduce the ratings of the equipment. Raised floor channels or concrete pads are often used to elevate distribution equipment above the floor, and thereby reduce the risk of contamination from water or leaking equipment (see illustration #166).

Distribution Equipment Clearances (cont'd)

NEC Clearances and Conditions			
Voltage to Ground Volts	**Working Space (feet) for Conditions***		
	(1)	**(2)**	**(3)**
0 - 150	3	3	3
151 - 600	3	3.5	4
601 - 2,500	3	4	5
2,501 - 9,000	4	5	6
9,001 - 25,000	5	6	9
25,001 - 75,000	6	8	10
above 75,000	8	10	12

Table #49 - Clearances (USA)

Conditions:

(1) Exposed live parts on one side and no live or grounded parts on the opposite side of the working space, or exposed live parts on both sides effectively guarded by suitable wood or other insulating materials. Insulated wire or insulated busbars operating at 300 V or below are not considered live parts.

(2) Exposed live parts on one side and grounded parts on the opposite side. Concrete, brick, or tile walls are considered as grounded surfaces.

(3) Exposed live parts and ungrounded parts on either side of the operator.

Distribution Equipment Clearances (cont'd)

Tables #49 and #50 Exception: Working space is not required in back of equipment such as deadfront switchboards or control assemblies having no renewable or adjustable parts, such as fuses or switches; and when all connections are accessible from locations other than the back.

CEC Clearances and Conditions	
Voltage to Ground Volts	**Working Space metres**
0 - 750	1.0
751 - 2,500	1.2
2,501 - 9,000	1.5
9,001 - 25,000	1.9
25,001 - 46,000	2.5
46,001 - 69,000	3.0
above 69,000	3.7

Table #50 - Clearances (Canada)

Metering Equipment

Metering provides a means of monitoring and measuring electrical circuit values such as voltage, current, power and energy. Utility companies calculate energy charges to customers by means of kilowatthour (kWh) meters and kVA demand meters. System voltages, currents and configuration (single phase, three-phase wye, etc.) determines which meter and metering equipment is needed.

Modern kWh meters are available in digital form, as shown in illustration #150A.

Most kWh meters are still of the electromechanical clock dial type as seen in illustration #150B. A kVA demand meter is shown in illustration #150C.

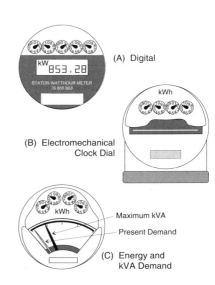

(A) Digital

(B) Electromechanical Clock Dial

(C) Energy and kVA Demand

Maximum kVA

Present Demand

Illustration #150 - Types of Metering Equipment

Metering Equipment (con't)

Kilowatthour meters measure only the true power consumed over a period of time. Most industrial and commercial customers have significant reactive loads and therefore kVA demand meters are installed. When a customer requires a large supply of electricity, even for a short period of time, the system must be designed to accommodate this requirement. The utility company passes this demand on to the customer by using the maximum kVA registered on the demand meter plus the actual energy consumed (kWh).

Electrical meters are limited in their voltage and current ratings. If voltages or currents above the rating of the meter are present, potential transformers (PT) as shown in illustration #151A, and current transformers (CT) as shown in illustration #151B, are used.

If currents in excess of 200 A and voltages in excess of 300 V are encountered, then instrument transformers are required.

Secondary CT current values are either 5 A or 1 A, and secondary PT voltage values are either 120 V or 69 V.

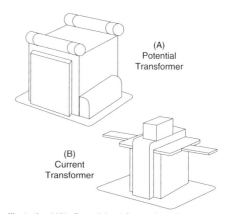

(A)
Potential Transformer

(B)
Current Transformer

Illustration #151 - Potential and Current Transformers

Metering Equipment (con't)

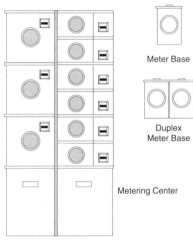

Meter Base

Duplex
Meter Base

Metering Center

Illustration #152 - Meter Bases

Note: Extreme caution must be used when using current transformers. The secondary of a current transformer must never be open circuited. When not using the secondary of a CT, short circuit the secondary windings.

Energy meters are required to be placed in meter bases. Meter bases are available in single, duplex or multiple base metering centers as shown in illustration #152.

Low Voltage Meter Connections

Electrical configurations and connections of meter bases vary and depend upon the voltages and number of phases present. Illustration #153A-E shows meter connections for some common low voltage systems.

Low Voltage Meter Connections (cont'd)

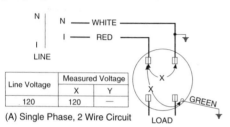

Line Voltage	Measured Voltage	
	X	Y
120	120	—

(A) Single Phase, 2 Wire Circuit

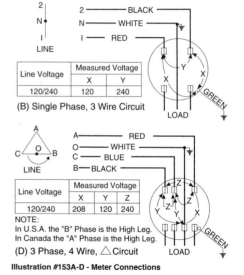

Line Voltage	Measured Voltage	
	X	Y
120/240	120	240

(B) Single Phase, 3 Wire Circuit

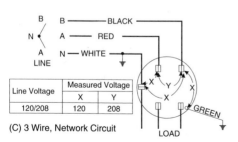

Line Voltage	Measured Voltage	
	X	Y
120/208	120	208

(C) 3 Wire, Network Circuit

Line Voltage	Measured Voltage		
	X	Y	Z
120/240	208	120	240

NOTE:
In U.S.A. the "B" Phase is the High Leg.
In Canada the "A" Phase is the High Leg.

(D) 3 Phase, 4 Wire, △ Circuit

Illustration #153A-D - Meter Connections

Low Voltage Meter Connections (cont'd)

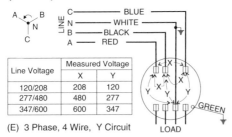

Line Voltage	Measured Voltage	
	X	Y
120/208	208	120
277/480	480	277
347/600	600	347

(E) 3 Phase, 4 Wire, Y Circuit

Illustration #153E - Meter Connections

Note: Configurations shown may not conform to local regulations. Confirm connections with local utility company.

Panelboards

Panelboards are an assembly of buses and connections, overcurrent devices, and control apparatus with or without switches for the control of light, heat, power or instrument circuits.

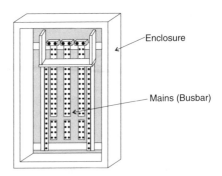

Illustration #154 - Panelboard and Busbars

Panelboards are designed to be placed in a cabinet or cutout box placed in or against a wall or partition and accessible from the front only. Panelboards are available for surface mounting and for flush or recessed mounting. Panelboards offer a convenient means of grouping circuit protective devices and load connections.

Panelboards (cont'd)

A panelboard consists of a set of copper or tin finished aluminum busbars, called mains, as shown in illustration #154. Individual branch circuits are tapped off from the mains through overcurrent devices which are usually molded case circuit breakers.

Panelboards are not designed nor approved as wireways or splicing boxes. Circuit conductors must terminate on approved overcurrent devices or on approved terminal blocks.

Note: Combination service/load center panelboards have a barrier between the service side and the load distribution side as shown in illustration #155. Panelboards without integral main overcurrent protection are called mains which means that the panelboard has only main lugs as shown in illustration #154.

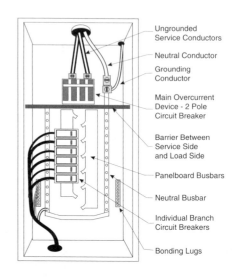

Ungrounded Service Conductors

Neutral Conductor

Grounding Conductor

Main Overcurrent Device - 2 Pole Circuit Breaker

Barrier Between Service Side and Load Side

Panelboard Busbars

Neutral Busbar

Individual Branch Circuit Breakers

Bonding Lugs

Illustration #155 - Panelboard With Barrier

Panelboards (cont'd)

Individual branch circuit breakers tap from the main busbars of the panelboard and are either of the stablock (inserted) type or of the bolted type as shown in illustration #156.

Overcurrent devices are available in single pole, two pole, or three pole units.

Panelboard overcurrent branch circuit devices are usually numbered and indexed as shown in illustration #157. The numbering arrangement used is consistent for most panelboards regardless of the amperage, voltage or number of phases.

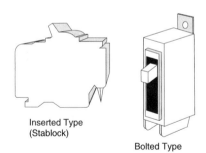

Inserted Type
(Stablock)

Bolted Type

Illustration #156 - Panelboard Branch Circuit Breakers

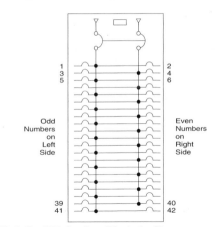

1
3
5

2
4
6

Odd
Numbers
on
Left
Side

Even
Numbers
on
Right
Side

39
41

40
42

Illustration #157 - Panelboard Numbering Diagram

Panelboards (cont'd)

The NEC articles 384-14 and 384-15 specify the classification of panelboards. The CEC does not classify panelboards. According to the NEC, a panelboard is classified as a **lighting and appliance panelboard** if not more than 42 positions are available and **over 10%** of the single-pole slots are rated at **30 A or less** and are used for branch circuits to lighting and appliance loads with neutral connections (illustration #158A).Lighting panelboards are available in ratings of 100 to 600 A, 600 V, single or three phase, and with busbar bracings of up to 65,000 A.

A **power panelboard** will have **no more than 10%** of its overcurrent devices rated at **30 A or less**, and protecting neutral connected loads. Power panels are used to supply subpanels or large 2 pole or 3 pole loads (see illustration #158B). They are available in ratings of 100 to 1600 A, 600 V, single or three phase, and with busbar bracings up to 100,000 A.

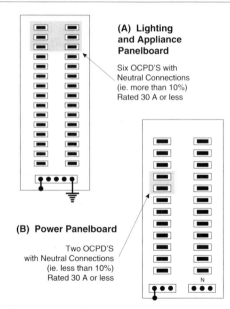

(A) Lighting and Appliance Panelboard

Six OCPD'S with Neutral Connections (ie. more than 10%) Rated 30 A or less

(B) Power Panelboard

Two OCPD'S with Neutral Connections (ie. less than 10%) Rated 30 A or less

Illustration #158 - NEC Panelboards

Motor Control Centers (MCC)

A motor control center (MCC) consists of a grouping of motor starters, controls, metering instruments, and may also include panelboards and programmable logic controllers (PLC), all housed within a common metal enclosure. MCCs may be arranged in straight-line, L-shaped, U-shaped or back-to-back configurations. The main bus as seen in illustration #159 usually runs horizontally across the top of the MCC.

The vertical bus, which has a lower amperage rating than the horizontal bus, supplies power to individual motor starters or equipment. Each vertical section of the MCC is sectionalized with starters of different sizes occupying a different number of compartments.

Compartments in each section of an MCC may be of the withdrawable type, thus allowing for quick replacement when equipment is not operating.

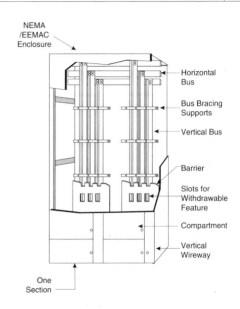

Illustration #159 - Motor Control Center

Motor Control Centers (cont'd)

Manufacturers have not standardized on the dimensions of MCC starters and their associated equipment. However, each manufacturer does publish all necessary statistics relevant to their equipment.

MCC Prewiring

MCCs are prewired in accordance with the buyers specifications. Two "Classes" of prewiring are available, and within each of these classes there are three "Types" of wiring. NEMA/EEMAC classifies MCC wiring as follows:

Class 1: No wiring by the manufacturer between compartments, and no interlocking between units or to remote mounted devices, and no control system engineering. Drawings of individual units are supplied.

Class 2: Prewiring by the manufacturer with interlocking and other control wiring completed between compartments of the MCC. Drawings illustrating the operation of the MCC are supplied by the manufacturer.

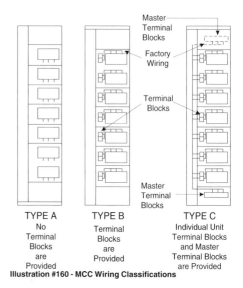

Illustration #160 - MCC Wiring Classifications

TYPE A
No Terminal Blocks are Provided

TYPE B
Terminal Blocks are Provided

TYPE C
Individual Unit Terminal Blocks and Master Terminal Blocks are Provided

Within the two wiring classes there are three wiring types as shown in illustration #160.

Motor Control Centers (cont'd)

Type A: No terminal blocks are provided with the MCC.

Type B: All connections within each compartment are made to terminal blocks by the manufacturer.

Type C: All connections are made to master terminal blocks located in the horizontal wiring trough either at the top or bottom of the MCC by the manufacturer.

Low voltage MCCs (600 V or less) are available in ratings of 600 A to 3000 A, with busbar bracings rated at 22,000, 42,000, 65,000 and 100,000 A symmetrical. Starter sizes range from NEMA/EEMAC #1 to #9 (#1-#6 common), (#7-#9 not very common). Motor starters are available in full voltage non-reversing, full voltage reversing, reduced voltage (autotransformer/wye-delta), two-speed single winding and two-speed two winding. MCCs are usually located indoors, or in a sheltered outdoor structure.

NEC Outdoor Motor Controller Enclosure Selection

Some protection against the following conditions:	Enclosure Type Number *						
	3	3R	3S	4	4X	6	6P
Incidental contact with the enclosed equipment	X	X	X	X	X	X	X
Rain, Snow and Sleet	X	X	X	X	X	X	X
Sleet**	-	-	X	-	-	-	-
Windblown dust	X	-	X	X	X	X	X
Hosedown	-	-	-	X	X	X	X
Corrosive Agents	-	-	-	-	X	-	X
Occasional Temporary Submersion	-	-	-	-	-	X	X
Occasional Prolonged Submersion	-	-	-	-	-	-	X

* Enclosure type number is marked on the motor controller enclosure.
** Mechanism is operable when covered with ice.

Table #51A - MCC Outdoor Enclosures

Motor Control Center (cont'd)

The NEC and CEC require that MCCs have enclosures that are suitable for the environment in which they are to be placed.

Tables #51A and #51B list NEC enclosure requirements for MCCs.

NEC Indoor Motor Controller Enclosure Selection											
Some protection against the following conditions:	Enclosure Type Number *										
	1	2	4	4X	5	6	6P	11	12	12K	13
Incidental contact with the enclosed equipment	X	X	X	X	X	X	X	X	X	X	X
Falling Dirt	X	X	X	X	X	X	X	X	X	X	X
Falling liquids and light splashing	-	X	X	X	X	X	X	X	X	X	X
Circulating dust, lint, fibers and flyings	-	-	X	X	-	X	X	-	X	X	X
Settling airborne dust, lint, fibers and flyings	-	-	X	X	X	X	X	-	X	X	X
Hosedown and splashing water	-	-	X	X	-	X	X	-	-	-	-
Oil and coolant seepage	-	-	-	-	-	-	-	-	X	X	X
Oil or coolant spraying or splashing	-	-	-	-	-	-	-	-	-	-	X
Corrosive Agents	-	-	-	X	-	-	X	X	-	-	-
Occasional Temporary Submersion	-	-	-	-	-	X	X	-	-	-	-
Occasional Prolonged Submersion	-	-	-	-	-	-	X	-	-	-	-
* Enclosure type number, except type number 1, is marked on the motor controller enclosure											

Table #51B - MCC Indoor Enclosures

Motor Control Center (cont'd)

Medium voltage (over 600 V to 18 kV)

motor controllers are built to different ANSI and IEC standards than low voltage MCCs. The following are medium voltage MCC ratings:

Voltages: 2.4 kV to 17.5 kV
Continuous currents: up to 4000 A
Interrupting ratings: 250 MVA to 1000 MVA
25 kA to 40 kA

Vacuum interrupter circuit breaker technology is commonly employed in modern medium voltage motor controllers. Each compartment has the option of containing withdrawable equipment, as shown in illustration #161.

MCC maintenance primarily involves keeping the enclosure clean and dry, and keeping electrical connections tight. Starter module doors are gasketed to stop dust and dirt from entering. These gaskets should be checked, and replaced if necessary, to maintain an effective barrier to contaminants.

Replace air filters, when used, on a regular basis.

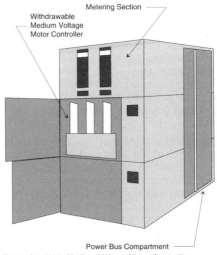

Illustration #161 - Medium Voltage Motor Controllers

Medium Voltage Motor Control Center (cont'd)

Vacuum or wipe the interior of the MCC on a regularly scheduled basis. When the MCC is scheduled for a periodic shutdown, check the tightness of buses and motor lead connections. The torquing specifications for bus hardware is the same as for metal-enclosed switchgear. Check the tightness of the mounting hardware for heater elements in the motor overload modules. Lubrication is not a concern for equipment located within an MCC.

Withdrawable units that are difficult to remove should be checked for alignment and/or distortion of the sliding assembly. When oxidants are being removed from copper buses, use aluminum oxide paper or a teflon scouring pad. Be very careful not to let the residual dust settle on other parts of the MCC. Never sand aluminum busbars because some of the aluminum plating might be removed.

Note: Worn or pitted motor starter contacts should not be reshaped or sanded, but should be replaced. Reshaping or sanding the contacts will remove the silver plating and sand or oxide particles may become embedded in the contact surface, resulting in an ineffective low resistance contact.

Programmable Logic Controllers (PLCs)

PLCs are used extensively in industry for a wide variety of applications. Demand for PLCs continues to grow as electrical equipment controlled and/or monitored by PLCs increases.

A PLC is a digital electronic device that uses programmable memory to store instructions for implementing specific operations such as logic, sequencing, timing, counting, and arithmetic operations to control machines and processes, and to provide interstation communications. A PLC provides an alternative to hardwired relays, counters, timers, sequencers and electrical interlocks.

PLC Motor Control (cont'd)

Some of the advantages to the use of PLCs are:

1. Eliminates costly rewiring
2. Increased flexibility
3. Centralized control
4. Adaptability for remote control and communications
5. Trouble shooting feedback
6. Faulted circuit or circuits may be bypassed quickly, thereby improving continuity of service
7. Lower costs for wire and wiring
8. Lower maintenance costs

Illustration #162 is a block diagram of a PLC system.

PLCs are programmed using ladder logic and Boolean statements, and employ numbering systems to assign functions, devices, and types of operations, as shown in illustration #163.

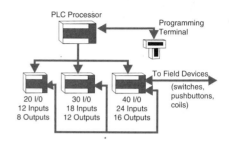

Illustration #162 - PLC System

Switchgear

Modern electrical distribution systems use metal-enclosed switchgear for the protection and control of feeder circuits, motors, lighting, generators, or other secondary switchgear. Switchgear is usually classified by its voltage rating and by its enclosure type.

Switchgear (cont'd)

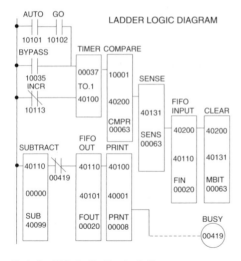

Illustration #163 - Ladder Type Logic Diagram

Two general types of switchgear are available:

a. open type
b. metal enclosed

Open type switchgear consists of distribution equipment located in a non-enclosed vertical housing with metering instruments and relays mounted on a front facing panel.

Located behind the panel are an assembly of buses, breakers, isolating switches, metering transformers, and cable connectors. Metal enclosed switchgear is much more common than open type switchgear.

Low voltage (600 V or less) switchgear are typically available in the following ratings:
Voltage rating: 208, 240, 480, 600 V AC; 125, 250 V (DC)
Continuous current ratings: 100 to 4000 A
Busbar bracing: 22,000 to 200,000 A

Switchgear (cont'd)

Medium voltage (2.3 kV to 35 kV) switchgear are typically available in the following ratings:

Voltage rating: 2.3, 5.0, 7.2, 13.8, 25, 34.5 kV
Continuous current ratings: 1200, 2000, 3000 A
Busbar bracing: 18,000 to 41,000 A

Switchgear Metal Enclosure

The metal enclosure of the switchgear serves to do the following:

1. support the internal components
2. separate personnel from electrical potentials
3. retard the accumulation of dust and dirt from entering the internal components
4. provide a fire barrier between adjacent cubicles, thus minimizing damage

Note: Electrical components are located in separate compartments to prevent failure of one component from affecting the continued operation of another. When doing modifications or maintenance to switchgear, the wiring from one compartment should not be routed through another compartment.

An example of an improper modification would be to pass control wiring through the power bus compartment without proper shielding or the protection of a metal conduit, or to lay the control wiring over the busbars. In the event of a severe failure, the potential exists for another system to fail. Control components are usually not placed in the same compartments with power components like breakers and disconnect switches, in order to reduce the risk of damaging control equipment in the event of a power components failure.

Switchgear (cont'd)

However, if an effective barrier exists within a compartment that contains a breaker or a disconnect switch, then panel mounted instruments may be located there. ANSI standards allow for the door of a drawout circuit breaker compartment to be used for mounting an instrument or relay if the circuit breaker has an appropriate front plate barrier. As wiring modifications are made to a piece of switchgear, barriers may be added to allow for sectionalizing or compartmentalizing. However, no barriers should be added in a switchgear assembly if ventilation will be blocked or clearances to live parts are too restrictive.

Metal enclosures are covered with a protective coating of paint. The paint has properties which will protect the metal and make it resistant to heat, oxidation and corrosion.

Classifications of Metal Enclosed Switchgear

Metal enclosed switchgear may be classified as being within one of the following groups:

1. Metal Enclosed:

a. Electrical components are contained in a single enclosure with minimal compartmentalization.

b. Circuit breakers may be fixed or be of the drawout type with a racking mechanism, as shown in illustration #164.

c. The main busbars do not have to be insulated, nor do they have to be segregated from the cable compartment.

2. Metal Clad:

a. The main interrupting device, which is usually a circuit breaker, is removable and/or withdrawable, and is so arranged that three positions are possible:
- connect position
- test position
- disconnect position

Metal Enclosed Switchgear (cont'd)

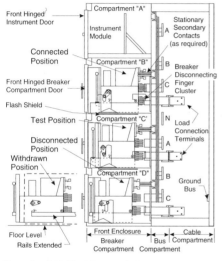

Illustration #164 - Metal Enclosed Switchgear

b. The circuit breaker, busbars, potential transformers and control power transformers are completely enclosed by grounded metal barriers, which have no intentional openings between compartments.

c. An inner barrier protects all or part of the circuit interrupting device to ensure that no primary circuit components are exposed when the unit door is opened, as seen in illustration #165.

d. All live parts are enclosed within grounded metal compartments.

e. Mechanical interlocks are provided to ensure a proper and safe operating sequence.

f. Automatic shutters cover primary circuit elements whenever removable devices such as circuit breakers are withdrawn.

g. Primary bus conductors and connections are insulated.

Metal Enclosed Switchgear (cont'd)

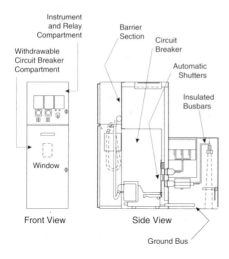

Instrument
and Relay
Compartment

Withdrawable
Circuit Breaker
Compartment

Barrier
Section

Circuit
Breaker

Automatic
Shutters

Insulated
Busbars

Window

Front View

Side View

Ground Bus

Illustration #165 - Metal Clad Switchgear

3. Arc Resistant Metal Clad:

This class of switchgear is based on an IEC standard.

a. All design features mentioned for metal clad switchgear are also present in arc resistant metal clad switchgear.

b. Allowances are made for internal over-pressures acting on covers, doors, inspection windows, etc.

c. Accessibility Type "A" enclosures are designed to be arc resistant at the front of the switchgear only.

d. Accessibility Type "B" enclosures are designed to be arc resistant at the front, back and sides.

e. Accessibility Type "C" enclosures are designed to be arc resistant at the front, back and sides, and between compartments within the same cell or an adjacent cell.

Metal Enclosed Switchgear (cont'd)

One method used to make metal clad switchgear arc resistant is to provide a safe means of venting and dispersing overpressures brought about by fault currents. Arc energy resulting from the arc developed under atmospheric air pressure will cause an internal overpressure and local overheating, resulting in mechanical and thermal stressing of the switchgear equipment. Materials within the switchgear may produce hot decomposition products, either gaseous or vaporous, which may be discharged to the outside of the enclosure.

As seen in illustration #166, venting flaps are placed in the roof of the switchgear. When arc energy results in the production of a severe overpressure, the venting flaps are blown open, thereby releasing the arc energy up instead of through the compartments or through the sides and doors. This design results in increased protection for personnel and equipment.

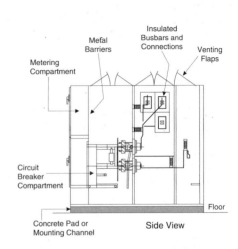

Illustration #166 - Arc Resistant Metal Clad Switchgear

Switchgear Maintenance
Note: The three most important points in switchgear maintenance are:
1. Keep the switchgear clean
2. Keep the switchgear dry
3. Keep the bus connections tight

Cleaning and tightening bus connections are performed during scheduled plant shutdowns when the switchgear can be made safe by de-energizing and grounding the power buses. The removing of dust from buses, connections, supports, insulators and enclosures can be accomplished most effectively with a vacuum cleaner.

Cloths moistened with water-free solvents are also effective in removing dust, dirt and other contaminants from the switchgear. Using compressed air to blow away dust and dirt may introduce problems such as embedding contaminants into the busbar insulation. Soap and water are effective on the most severe cases of oil or carbon contamination (such as after a fire).

Ensure that all equipment is thoroughly dry before restoring into service.

Check the tightness of bolted connections in the switchgear, especially the bolted connections of busbars. Table #52 provides bolt torques for bolt sizes and materials used on bus materials.

Bolt Torque for Bus Connections					
	Torque in foot-pounds for bolt size				
Bolt Material	**.25- 20**	**.31- 18**	**.38- 16**	**.50- 13**	**.62- 11**
High Strength Steel	5	12	20	50	95
Silicon Bronze	5	10	15	40	55
Note: Bolt sizes are expressed in inches and threads are given in number per inch					

Table #52 - Busbar Bolt Torques

Switchgear Maintenance (cont'd)

Checks should be made for alignment of mechanical rails and moving parts to ensure that equipment is properly aligned to their bus connections. Mechanical maintenance consists mainly of keeping these parts clean and in adjustment.

Testing of the breakers, buses, and other switchgear components is required periodically. Insulation tests using either a megohmmeter for low voltage switchgear, or high potential AC or high potential DC for medium voltage switchgear, will reveal if any insulation weaknesses are present.

The following list of maintenance checks are recommended to be undertaken by qualified electrical maintenance staff on metal switchgear:

Annually:

1. *Trip and close each breaker* electrically in the "test position" and test all the protective relays* and switches associated with a particular breaker.

*   A protective relay uses current and/or voltage signals to process information. When this information falls within a preset range of values, which are predetermined to represent a fault condition, the protective relay produces an output action that results in a breaker trip. Protective relays are either electro-mechanical, electronic, or microprocessor based, as shown in illustration #167.

2. *Open cover plates and clean enclosures* with a soft brush and a vacuum cleaner, then hand clean the enclosure with soft rags that have been moistened with a non-water solvent.

Switchgear Maintenance (cont'd)

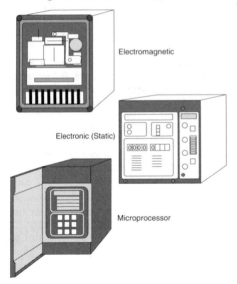

Electromagnetic

Electronic (Static)

Microprocessor

Illustration #167 - Protective Relays

3. *Megohmmeter test the busbars*, phase to phase, and phase to ground. Record the results and compare with the previous readings.

4. *Check all structural hardware* for tightness.

5. *Check all switchgear grounds*. Record resistivity values of the switchgear ground and compare to previous readings.

6. *Renew lubrication on primary bushings*, secondary fingers and racking gear.

7. *Operate draw-out and racking mechanisms*. Check shutters, interlocks and limit switches.

8. *Inspect breakers*, checking their positioning and primary and secondary contacts.

Switchgear Maintenance (cont'd)

9. *Service batteries and battery charger.* Check and record the charging rate and the voltage level.

10. *Check the instrument transformers* and the connections for wear, fatigue and overheating. Check the fuses of the potential transformers.

11. *Check protective relays* and panel devices. Remove covers, blow out dust and dirt with low pressure air or nitrogen. Check that all indicating lights and sockets are operating properly.

Every three years:

1. *Clean all busbars and supporting material,* and check for signs of tracking, corona (look for a white powder on or around busbars) or overheating.

2. *Check all secondary wiring* for tightness.

3. *Conduct megohmmeter tests of controls and relay circuits* to ground.

4. *Check the enclosure* for leaks, door operations, floor levelness, ventilation, space heaters and lighting.

5. *Perform a dielectric insulation test* on the busbars and record the values.

SECTION SEVEN
MOTORS

Motor Types

Electric motors are manufactured in a number of different types and may be divided mainly into three groups:

1. DC Motors:

Shunt connected, series connected, cumulatively compound connected and permanent magnet.

2. AC Polyphase Motors:

Squirrel cage induction, wound rotor and synchronous.

3. AC Single Phase Motors:

Split phase resistor start, split phase capacitor start, split phase permanent capacitor, shaded pole, series (universal) and hysteresis.

Nameplate Data

The rating of an electric motor includes the following information: service classification (general purpose, definite purpose, special purpose), voltage, full load current and speed.

It also includes the number of phases and frequency (if alternating current), full load horsepower or full load kW, power factor (when applicable), service factor, insulation class, locked rotor kVA code, rotor torque design code (if alternating current), duty classification, frame and enclosure type.

A motor nameplate, as in illustration #168, shows the ratings of the motor and may also include a winding connection diagram.

Motor Nameplate Terms

Power Rating: The full load horsepower or full load kW rating stamped on the nameplate of a motor refers to the allowable load that a motor may carry without injury to any part of the motor. The power rating of a motor refers to the value of output power that a motor will produce at the shaft without overheating.

Service Factor: The service factor of an alternating current motor is a multiplier indicating a permissible loading above nameplate values.

Motor Nameplate Terms (cont'd)

Multiplying nameplate power by the service factor reflects how much the motor may be overloaded, preferably for brief periods. Common service factors for AC motors are 1.0, 1.1 and 1.15.

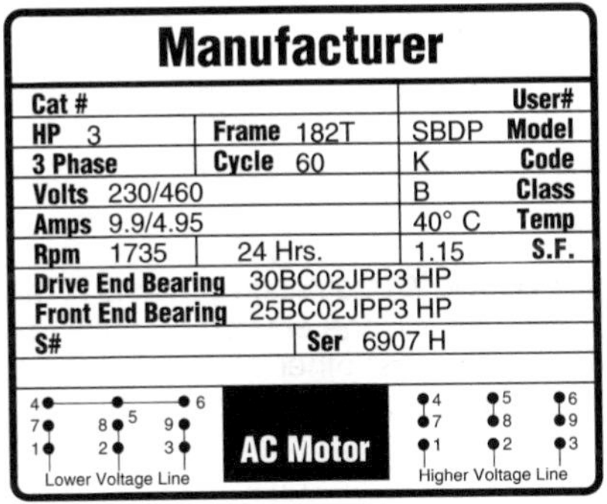
Manufacturer
Cat #
User#
HP 3
Frame 182T
SBDP Model
3 Phase
Cycle 60
K Code
Volts 230/460
B Class
Amps 9.9/4.95
40° C Temp
Rpm 1735
24 Hrs.
1.15 S.F.
Drive End Bearing 30BC02JPP3 HP
Front End Bearing 25BC02JPP3 HP
S#
Ser 6907 H
4 5 6
7 8 9
1 2 3
Lower Voltage Line
AC Motor
4 5 6
7 8 9
1 2 3
Higher Voltage Line

Illustration #168 - Motor Nameplate Example

Note: Whenever a motor is operated above its rated power value, the temperature of the motor windings will increase.

For every 18°F (10°C) rise in insulation temperature above rated values the life of the motor is cut in half. It is strongly recommended that motors be operated within the range of their ratings and overloaded only for brief periods.

Insulation Class: Motor insulations are divided into temperature classes. Four classes of insulation are used: Class A, Class B, Class F and Class H. Table #52 lists the temperature limitations of each class and illustration #169 graphs the life expectancy of these insulations as compared to temperature.

Class A (105°C) insulation consists of materials such as cotton, silk and paper which is suitably impregnated, coated, or immersed in a dielectric liquid such as oil.

Motor Nameplate Terms (cont'd)

Motor Insulation Classes				
	A	B	F	H
Ambient Temp.°C	40	40	40	40
Allowable Temperature Rise °C	60	80	105	125
Hot Spot Allowance °C	5	10	10	15
Hot Spot Temp.°C	105	130	155	180

Table #52 - Motor Insulation Classes

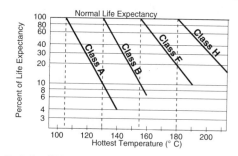

Illustration #169 - Insulation Class Graph

Class B (130°C) insulation consists of materials such as mica, glass, fiber, asbestos or other materials with suitable bonding substances.

Class F (155°C) insulation consists of materials such as mica, glass fiber, asbestos or other materials with suitable bonding substances such as epoxy enamels.

Class H (180°C) insulation consists of materials such as silicone elastomer, mica, glass fiber, asbestos or other materials with suitable bonding substances such as silicone resin.

Motor Nameplate Terms (cont'd)

Locked kVA Code: When motors are started, held in a locked rotor condition, or stalled, they will draw line currents whose magnitudes are dependent upon their rotor construction. As starting current or stalled rotor current has a direct bearing on the settings of control protective devices, a code has been developed that is used in calculating the starting kVA per horsepower of the motor. Motor code letters and locked rotor kVA values are listed in table #53.

Rotor Torque Design (AC squirrel cage motors): The torque characteristics of motors are represented by NEMA/EEMAC design classes; design A, design B, design C, design D and design F. The values assigned to these design classes are summarized in table #54 (each design class represents a different rotor design, and the values shown are based on full voltage starting).

Code Letters for Locked Rotors	
Code Letter	kVA/Horsepower with Locked Rotor
A	0 - 3.14
B	3.15 - 3.54
C	3.55 - 3.99
D	4.0 - 4.49
E	4.5 - 4.99
F	5.0 - 5.59
G	5.6 - 6.29
H	6.3 - 7.09
J	7.1 - 7.99
K	8.0 - 8.99
L	9.0 - 9.99
M	10.0 - 11.19
N	11.2 - 12.49
P	12.5 - 13.99
R	14.0 - 15.99
S	16.0 - 17.99
T	18.0 - 19.99
U	20.0 - 22.39
V	22.4 and up

Table #53 - Locked Rotor kVA Values

Motor Nameplate Terms (cont'd)

NEMA/EEMAC Design Class	Full-Load Speed % Regulation	Start Torque x Rated Torque	Start Current x Rated	Characteristic Name
A	2 - 5	1.5 - 1.75	5 - 7	Normal
B	3 - 5	1.4 - 1.6	4.5 - 5	General Purpose
C	4 - 5	2 - 2.5	3.5 - 5	High Torque, Double Cage
D	5-8 8-13	Up to 3	3 - 8	High Torque, High Resistance
F	Over 5	1.25	2 - 4	Low Torque, Double Cage, Low Starting Current

Table #54 - Motor Rotor Design Class Characteristics

Duty Classification: Motors are classified by the duty of their load cycles. A motor may have a load duty as being either continuous duty, short-time duty, intermittent duty, periodic duty or varying duty. Both the NEC and the CEC require that motor circuit conductors be sized in accordance with the duty class of the motor. Within some classes there may be time ratings of 5, 15, 30 or 60 minutes.

The base upon which the time rating is referred to should be obtained from the motor manufacturer. As an example, a 15 minute rated motor operating for a full 15 minutes may have to rest for 45 minutes before operating again, and a 30 minute rated motor operating for a full 30 minutes may have to rest for 30 minutes before operating again.

Motor Nameplate Terms (cont'd)

Enclosure Type: NEMA/EEMAC have classified motor enclosure types. The following list of enclosures includes many but not all of the enclosure types:

a. **Open enclosure** - An open end frame structure that permits maximum air circulation for ventilation. This construction is designed to prevent falling objects from making contact with electrically live or moving parts.

b. **Drip-proof enclosure** - An open machine construction in which the ventilating openings are so constructed that successful operation is not interfered with when drops of liquid or solid particles strike or enter the enclosure at any angle from 0 to +/- 15 degrees downward from the vertical plane.

c. **Splash-proof enclosure** - An open machine in which the ventilating openings are so constructed that successful operation is not interfered with when drops of liquid or solid

particles strike or enter the enclosure at any angle not greater than +/- 100 degrees downward from the vertical plane.

d. **Guarded enclosure** - The enclosure is so arranged that no accidental or intentional object can penetrate.

e. **Weather-protected Type 1** - An open machine with its ventilating passages so constructed as to minimize the entrance of rain, snow and air-borne particles to the electric parts.

f. **Weather-protected Type 2** - Similar to Type 1, except for the arrangement of intake and discharge ventilating passages. High velocity air and air-borne particles that are blown into the machine by storms or high winds can be discharged without entering the internal ventilating passages of the motor that lead directly to the electrical parts of the machine.

Motor Nameplate Terms (cont'd)

g. *Totally enclosed nonventilated (TENV)*

A totally enclosed motor which is not equipped for cooling by means external to the enclosing parts.

h. *Totally enclosed fan cooled (TEFC)* - A totally enclosed motor which is equipped for exterior cooling by means of a fan or fans integral with the motor but external to the enclosing parts.

i. *Explosion proof motor* - A totally enclosed motor, with an enclosure designed and constructed to withstand and contain an explosion of a specified gas or vapor which may have occurred within it. The enclosure prevents the ignition of the gases or vapors surrounding the machine and contains sparks, flashes or explosions which may have occurred within the motor casing.

j. *Dust ignition proof* - A totally enclosed motor with an enclosure that is designed and constructed in a manner which will exclude ignitable amounts of dust which might affect performance or rating. It will not permit arcs, sparks, or heat otherwise generated or liberated inside of the enclosure to cause ignition of exterior accumulations or atmospheric suspensions of a specific dust on or in the vicinity of the motor.

k. *Water proof enclosure* - A totally enclosed motor so constructed that it will exclude water from entering when applied in the form of a stream coming from a hose.

Frame Size: NEMA/EEMAC have a standardized method of sizing the frames of motors. These frame sizes may be visualized as diameter series with a few standardized lengths within each diameter series.

Motor Nameplate Terms (cont'd)

Frame sizes are used to obtain frame dimensions such as, shaft height, shaft length, shaft diameter, distances between mounting bolts, overall length of motor, and the diameter of the motor. Older motors are U frame (before 1964), and newer motors are T frame (after 1964) for motors up to 250 HP.

Voltage of the Motor: NEMA/EEMAC have standard (nominal) voltage ratings for motors as follows:

a. AC single phase line to line voltages - 115, 230, 460 V

b. AC polyphase line to line voltages - 110, 208, 220, 440, 550, 2300, 4000, 4600, 6600 V

c. DC voltages - 120, 240, 550 V

For DC motors the speed and operating current will change when the voltage rises or falls below its rated value, as noted in tables #55A and #55B.

Table #56 lists the changes that occur in AC T frame motors when the supply voltage varies above or below rated values.

Note: The values shown in table #56 are based on balanced line voltages.

Note: The motor nameplate voltage should always be checked against the system voltage that the motor is being connected to.

Motor Nameplate Terms (cont'd)

Characteristics of Shunt Wound DC Motors with Voltage Variations			
	120% voltage	110% voltage	90% voltage
Starting/Max. running torque	Increase 30%	Increase 15%	Decrease 16%
Full Load Speed	110%	105%	95%
Efficiency - full load	Slight increase	Slight increase	Slight decrease
Efficiency - 3/4 load	No change	No change	No change
Efficiency - 1/2 load	Slight decrease	Slight decrease	Slight increase
Full Load current	Decrease 17%	Decrease 8.5%	Increase 11.5%
Temperature rise - full load	Main field increases. Commutating field and armature decreases.	Main field increases. Commutating field and armature decreases.	Main field decreases. Commutating field and armature increases.
Maximum overload capacity	Increase 30%	Increase 15%	Decrease 16%
Magnetic noise	Slight increase	Slight increase	Slight decrease

Table #55A - DC Motor Voltage Variations

Motor Nameplate Terms (cont'd)

Characteristics of Compound Wound DC Motors with Voltage Variations			
	120% voltage	110% voltage	90% voltage
Starting/Max. running torque	Increase 30%	Increase 15%	Decrease 16%
Full Load Speed	112%	106%	94%
Efficiency - full load	Slight increase	Slight increase	Slight decrease
Efficiency - 3/4 load	No change	No change	No change
Efficiency - 1/2 load	Slight decrease	Slight decrease	Slight increase
Full Load current	Decrease 17%	Decrease 8.5%	Increase 11.5%
Temperature rise - full load	Main field increases. Commutating field and armature decreases.	Main field increases. Commutating field and armature decreases.	Main field decreases. Commutating field and armature increases.
Maximum overload capacity	Increase 30%	Increase 15%	Decrease 16%
Magnetic noise	Slight increase	Slight increase	Slight decrease

Table #55B - DC Motor Voltage Variations

Motor Nameplate Terms (cont'd)

Characteristics of T-Frame Motors with Voltage Variations			
	90% voltage	110% voltage	Function of Voltage
Starting/Max. running torque	Decrease 19%	Increase 21%	Voltage^2
Percent Slip	Increase 20 - 30%	Decrease 15 - 20%	$1/(Voltage)^2$
Full Load Speed	Slight decrease	Slight increase	Synchronous speed slip
Efficiency - full load	Decrease 0 - 2%	Decrease 0 - 3%	
Efficiency - 3/4 load	No change	No change/ slight decrease	
Efficiency - 1/2 load	Increase 0 - 1%	Decrease 0 - 5%	
Power factor - full load	Increase 1 - 7%	Decrease 5 - 15%	
Power factor - 3/4 load	Increase 2 - 7%	Decrease 5 - 15%	
Power factor - 1/2 load	Increase 3 - 10%	Decrease 10 - 20%	
Full Load current	Increase 5 - 10%	*Decrease 5 - 10%	
Starting current	Decrease approx.10%	Increase approx.10%	Voltage
Temperature rise - full load	Increase 10 - 15%	Increase 2 - 15%	
Maximum overload capacity	Decrease 19%	Increase 21%	Voltage^2
Magnetic noise	Slight decrease	Slight increase	

Note*: High Efficiency T-Frame motors will experience a current rise

Table #56 - T-Frame Motor Voltage Variations

Motor Nameplate Terms (cont'd)

Frequency: Applies only to AC motors and is 60 Hz in North America .

If the frequency to the motor should vary, the behavior of the motor changes as noted in table #57.

Characteristics of AC Motors with Frequency Variations			
	105% frequency	95% frequency	Function of Frequency
Starting/Max. running torque	Decrease 10%	Increase 11%	$1/(Frequency)^2$
Synchronous speed	Increase 5%	Decrease 5%	Frequency
Percent Slip	No change	No change	
Full Load Speed	Increase 5%	Decrease 5%	Nearly direct
Efficiency - full load	Slight increase	Slight decrease	
Efficiency - 3/4 load	Slight increase	Slight decrease	
Efficiency - 1/2 load	Slight increase	Slight decrease	
Power factor - full load	Slight increase	Slight decrease	
Power factor - 3/4 load	Slight increase	Slight decrease	
Power factor - 1/2 load	Slight increase	Slight decrease	
Full Load current	Slight decrease	Slight increase	
Starting current	Decrease 5 - 6%	Increase 5 - 6%	1/ Frequency
Temperature rise - full load	Slight decrease	Slight increase	
Maximum overload capacity	Slight decrease	Slight increase	
Magnetic noise	Slight decrease	Slight increase	

Table #57 - Frequency Variations

Motor Nameplate Terms (cont'd)

Speed: NEMA/EEMAC speed standards are 3500, 2500, 1750, 1150, 850, 650, 500, 400 and 300 RPM for DC motors and 3600, 1800, 1200, 900, 720 and 600 RPM for standard T frame AC motors.

HP	DC Motors - Full Load Current (Amps)					
	Armature Voltage Rating					
	90 V	120 V	180 V	240 V	500 V	550 V
1/4	4.0	3.1	2.0	1.6		
1/3	5.2	4.1	2.6	2.0		
1/2	6.8	5.4	3.4	2.7		
3/4	9.6	7.6	4.8	3.8		
1	12.2	9.5	6.1	4.7		
1 1/2		13.2	8.3	6.6		
2		17	10.8	8.5		
3		25	16	12.2		
5		40	27	20		
7 1/2		58		29	13.6	12.2
10		76		38	18	16
15				55	27	24
20				72	34	31
25				89	43	38
30				106	51	46
40				140	67	61
50				173	83	75
60				206	99	90
75				255	123	111
100				341	164	148
125				425	205	185
150				506	246	222
200				675	330	294

Table #58 - DC Motors Full Load Current Ratings

Motor Nameplate Terms (cont'd)

HP	Three phase AC Motors - Full Load Current (Amps)										
	Induction Type Squirrel Cage/Wound Rotor							Synchronous Type Unity Power Factor			
	115 V	200 V	208 V	230 V	460 V	575 V	2300V	230 V	460 V	575 V	2300V
1/2	4.0	2.3	2.2	2.0	1.0	0.8					
3/4	5.6	3.2	3.1	2.8	1.4	1.1					
1	7.2	4.1	4.0	3.6	1.8	1.4					
1 1/2	10.4	6.0	5.7	5.2	2.6	2.1					
2	13.6	7.8	7.5	6.8	3.4	2.7					
3		11.0	10.6	9.6	4.8	3.9					
5		17.5	16.7	15.2	7.6	6.1					
7 1/2		25.3	24.2	22	11	9					
10		32.2	30.8	28	14	11					
15		48.3	46.2	42	21	17					
20		62.1	59.4	54	27	22					
25		78.2	74.8	68	34	27		53	26	21	
30		92.0	88.0	80	40	32		63	32	26	
40		119.6	114.4	104	52	41		83	41	33	
50		149.5	143.0	130	65	52		104	52	42	
60		177.1	169.4	154	77	62	16	123	61	49	12
75		220.8	211.2	192	96	77	20	155	78	62	15
100		285.2	272.8	248	124	99	26	202	101	81	20
125		358.8	343.2	312	156	125	31	253	126	101	25
150		414	396	360	180	144	37	302	151	121	30
200		552	528	480	240	192	49	400	201	161	40

Table #59 - AC Three Phase Full Load Motor Current Ratings

Motor Nameplate Terms (cont'd)

Current Rating: The full load current rating of a motor refers to the current that the motor will be drawing when operating at rated speed and producing rated output power. Motors with the same power and voltage ratings may not necessarily have the same full load current values.

Nameplates or manufacturers' published values should always be consulted when determining the full load current of a motor. Table #58 for DC motors, table #59 for AC three phase motors, and table #60 for single phase motors, provide generic values of motor currents and are for reference purposes only.

AC Single Phase Motors				
HP	115 V	200 V	208 V	230 V
1/6	4.4	2.5	2.4	2.2
1/4	5.8	3.3	3.2	2.9
1/3	7.2	4.1	4.0	3.6
1/2	9.8	5.6	5.4	4.9
3/4	13.8	7.9	7.6	6.9
1	16	9.2	8.8	8
1 1/2	20	11.5	11.0	10
2	24	13.8	13.2	12
3	34	19.6	18.7	17
5	56	33.2	30.8	28
7 1/2	80	46.0	44	40
10	100	57.5	55	50

Table #60 - AC Single Phase Full Load Motor Current Ratings

DC Motors

There are hundreds of millions of direct current (DC) motors in service in North America. Almost all automotive motors are DC, from the starting motor to all the accessory motors such as power seats, power roofs and power trunks.

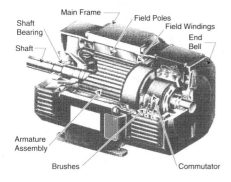

Illustration #170 - Typical DC Motor

DC traction motors are used in the railroad industry (newer locomotives now have AC motors), in elevators and in rapid transit urban trains. The speed control of DC motors is so precise that often DC motors are used as the drive for delicate laboratory instruments. Direct current motors are manufactured from the millihorsepower range to several hundred horsepower, and are available in voltages ranging from 1.5 V for millihorsepower motors to 6, 12, 24, 48, 120, 240, 550 and even 750 V.

DC Motor Construction

A DC motor (see illustration #170) principally consists of a main frame, field poles, field pole windings, armature structure, armature windings, commutator, brushes, shaft bearings and an end bell structure.

a. **The main frame** is made of carbon steel and serves as both a stationary support structure and as a return path for all the circulating magnetic flux that passes from the field poles to the armature.

DC Motors (cont'd)

b. **The field poles** are made of laminated steel alloys and serve as both the support structure for the field windings and as a magnetic path.

c. **The shunt field** consists of many turns of fine copper wire wrapped around the field poles. It is always connected in parallel with the armature winding. **The series field** winding consists of a few turns of heavy copper wire wrapped around the field poles and is always connected in series with the armature winding. Motors that have both a shunt and series field are called compound connected.

d. **The armature structure** consists of a stack of alloy steel discs assembled on to a shaft. The armature structure serves a dual purpose of providing support for the armature windings and also providing a path for magnetic fluxes.

e. **The armature windings** are placed within the notches of the armature structure and are designed to carry the large motor currents.

f. **The commutator** is a rotary rectifier or switch constructed of copper wedges separated by mica insulators.

g. **Carbon brushes** are used to make a low resistance contact with the commutator.

h. **The entire armature assembly** including the commutator is mounted on a **shaft** and supported on each end of the main frame by a shaft bearing assembly called end bells.

DC Motor Principles/Connections

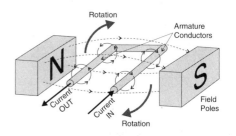

Illustration #171 - DC Motor Principles

DC Motors (cont'd)

Torque is rotational energy and is defined as a force acting through a distance that tends to cause an object to rotate. A DC motor produces torque through the interaction of the magnetic field of the rotor and the magnetic field of the armature circuit, as shown in illustration #171.

A hard wired electrical connection is made to each of the windings, and therefore currents flowing into the field and into the armature windings do so by conduction. The amount of magnetic flux developed by each winding depends on the amount of current that flows through each winding. The amount of torque developed depends on the amount of magnetic flux produced by each of the windings.

The amount of torque produced may be expressed as:

$$T \approx I_{ARMATURE} \times I_{FIELD}$$

There are three common wiring connections shown in illustration #172: shunt, series and compound (permanent magnet DC motors are an exception to these three connections and do not make use of a field winding as the field poles have a permanent magnetic orientation). Each of these wiring connections will produce a different operating characteristic.

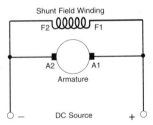

Illustration #172A - DC Shunt Connection

DC Motors (cont'd)

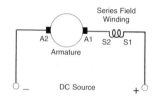

Illustration #172B - DC Series Connection

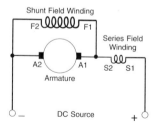

Illustration #172C - DC Compound Connection

DC Motor Operating Characteristics

Each of the three connection types behaves differently to changing load conditions. For operating characteristics refer to illustration #173A-D.

a. **Shunt motors** have the characteristic of developing low starting torques but good speed regulation. Typical applications include machine tools, automotive motors and applications where constant speed at any load is important. If the shunt field should accidentally be open circuited, large armature currents will develop, and if operating at no load, the armature speed will rise very quickly and may reach a level where the centrifugal forces may cause the motor to self destruct.

DC Motors (cont'd)

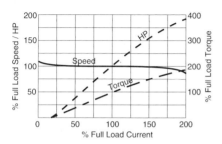

Illustration #173A - Shunt Motor

b. **Series motors** develop high starting torques with poor speed regulation. The speed of a series motor varies greatly depending on the load. Under no load the speed of the motor can be six times the normal operating speed and could produce centrifugal forces that will destroy the motor.

Series motors should either be direct coupled or geared to their load; they should never be belt connected to their load or operated at no load. Typical applications for series motors are traction motors for railroads or large earth moving equipment, rapid transit systems, service trucks like forklifts and electric powered carts.

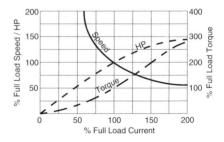

Illustration #173B - Series Motor

DC Motors (cont'd)

c. **Compound cumulatively** (series field and shunt field are aiding each other) connected motors have the characteristic of developing good starting torques and may have excellent speed regulation depending on their degree of compounding. By adding or subtracting series field winding turns the motor may be over, under or flat compounded.

Typical applications for compound motors are operating conveyors, elevators, air compressors, hoisting equipment, paper cutters, printing presses and centrifugal pumps.

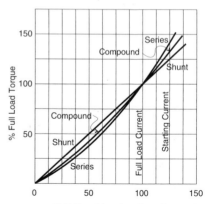

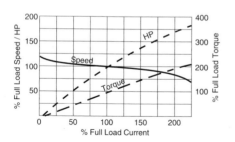

Illustration #173C - DC Compound Motor

Illustration #173D - DC Torque Variations

DC Motors (cont'd)
Control of DC Motors

A DC motor at rest will offer very little resistance to the flow of current. If rated voltage were applied to the motor, large currents would flow. To reduce these high starting currents a motor starter that employs some means of soft starting is always used. The earliest starters were the three point and four point manual starters (resistance was inserted into the armature circuit at start and gradually removed by the wiping action of a manually actuated sliding resistor). Modern starters use either automatic definite time electromagnetic devices as shown in illustration #174, or solid state starters.

In illustration #174, a field loss relay is used to sense that shunt field current is present before the armature circuit is turned on. Current limiting resistors, sequentially removed from the armature circuit, will limit the inrush current and thus allow for a soft start.

Timing relays are fully adjustable thereby allowing for the customizing of the acceleration time of the motor. The thyrite resistor serves to absorb high inductively generated voltages that are produced by the shunt field when the main overcurrent device or disconnect switch opens. This type of a control circuit no longer depends on the judgement of an operator, as was the case with the older three point or four point starters.

The starter in illustration #175 is essentially the same as the one shown in illustration #174 except that it contains a forward and reversing contactor with a mechanical interlock to prevent an armature short circuit and electrical interlocks to prevent coil burnout.

DC Motors (cont'd)

Note: Reversing a DC motor is accomplished by reversing either the field winding or the armature winding (reversing the armature winding is most common).

If both windings are reversed at the same time, the direction of the motor will remain the same.

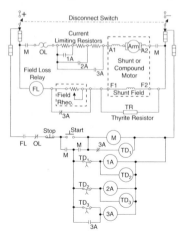

Illustration #174 - Automatic Definite Time DC Motor Starter

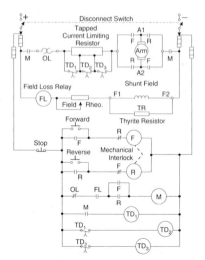

Illustration #175 - Forward & Reversing DC Motor Starter

DC Motors (cont'd)
Stopping a DC Motor

A DC motor may be slowed down or brought to a quick stop by using one of three electrical means:

a. dynamic braking

b. regenerative braking

c. plugging

Dynamic Braking: The motor is momentarily converted to a generator by inserting a resistor in parallel with the armature circuit when the stop button is pushed. The still turning motor is now acting like a generator, and the energy it produces is dissipated through the resistive element as heat. With this type of control, the motor cannot be brought to a full stop, it can only be used to slow the motor down. Completely stopping the dynamically braked motor requires either the application of a mechanical brake or an actual reversal of the motor.

Regenerative Braking: This involves the same principles as dynamic braking with the exception of not using resistors. The still turning motor, now acting like a generator, will have its energy pumped back into the system that is supplying it. This is only possible if the motor is now turning faster than when it was motoring. An example would be a diesel locomotive that is being used in a mountain grade service. The power developed by the now descending train is transferred into other electrical loads that are essential to the train's operations.

Plugging: If a motor requires sudden or repeated sudden stops, a plugging control switch may be employed. Plugging involves the reversal of the armature circuit polarity. At the moment of reversal, the torque direction is reversed and the rotational speed decreases to zero.

DC Motors (cont'd)
Speed Control of DC Motors

The speed of a DC motor is dependent upon two factors:

1. Armature voltage ($S \approx V_{ARMATURE}$)

2. Field current ($S \approx 1/I_{FIELD}$)

There are four general methods of controlling the speed of a DC motor:

1. **Field control** - Varying the amount of field current will alter the speed of the motor. The higher the field current the lower the speed of the motor and vice versa. Practical speed ranges are from 25% to 125% of rated speed. The action of field current control does not result in a linear stepless change in speed, and therefore this method of speed control is not the best.

2. **Armature resistance control** - A resistor is inserted in series with the armature circuit.

As a result the armature voltage will change, and thus the speed of the motor will change. The armature resistors are high wattage units that will produce heat and lower the efficiency of the system. Speed control and ranges vary depending on the values of resistance in the circuit. Speed ranges from 25% to 100% of rated speed are possible.

3. **Series and shunt armature resistance control** - Consists of the combination of (1) and (2). This results in a more complicated and not necessarily desirable arrangement with higher inefficiencies.

4. **Armature voltage control** - By varying the armature voltage stepless speed control is accomplished. This is the preferred way of controlling the speed of a DC motor. The range of speeds available are from 0% to over 100% of the rated speed of the motor. Solid state speed controllers provide stepless speed control, and they incorporate all of the safety features of a conventional automatic starter.

AC Polyphase Motors

The two most common types of polyphase motors are:

1. Two phase (voltages are displaced by 90 degrees and the voltage between lines is equal to $\sqrt{2}$ times phase voltage).

2. Three phase (voltages are displaced by 120 electrical degrees and voltage between lines is equal to $\sqrt{3}$ times phase voltage).

Three phase motors are the most common type in industry. They range in size from fractional horsepower to over 100,000 horsepower. Their simple design, reliability, ruggedness, low maintenance, good efficiency, lower cost than direct current motors of equal rating, and ability to handle temporary overloads has led to their widespread popularity. Polyphase motors are used in everything from machine tools in a shop or factory to driving compressors and fans for air handling systems and process pumps in industrial plants.

There are three groups of AC three phase motors:

- squirrel cage induction
- wound rotor
- synchronous

Squirrel Cage Induction Motor Construction

A polyphase motor, as shown in illustration #176, consists of a main frame or enclosure, a stator core, stator windings, squirrel cage rotor, shaft, shaft bearings, terminal box and end bells.

a. **The main frame/enclosure** assembly serves a dual purpose of housing the wound stator and of providing mechanical protection for the motor assembly.

b. **The stator core** is an assembly of thin laminations stamped from silicon-alloy sheet steel (silicon steel minimizes hysteresis losses and eddy currents are minimized by coating the laminations with an oxide or varnish).

AC Polyphase Motors (cont'd)

c. **Insulated coils (stator windings)**, are set in slots within the stator core. The stator windings are connected in a series or parallel arrangement to form phase groups, and these phase groups are then connected in either a wye or delta configuration depending on voltage and current requirements.

d. **The squirrel cage rotor**, as shown in illustration #177, consists of a slotted core of laminated steel into which molten aluminum or copper is cast to form the rotor conductors, the end rings and the fan blades. There is no insulation between the iron core and the aluminum conductors because the induced voltages are very low and the reduced currents will choose to flow through the low impedance aluminum conductors. The rotor is often skewed, as shown in illustration #177, primarily to minimize the noises due to magnetizing forces (minimize motor growl) and also to smooth out torque variations.

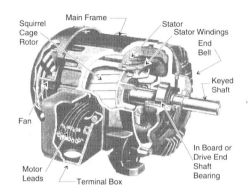

Illustration #176 - Polyphase Motor Construction

The squirrel cage rotor is pressed on to a steel shaft. The steel shaft in turn is fitted on to end bearings that are located on the end bell assembly.

AC Polyphase Motors (cont'd)

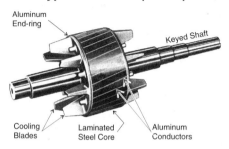

Illustration #177 - Squirrel Cage Rotor

Squirrel Cage Induction Motor Operating Principle

There are no physical electrical connections between the stator and the rotor of an induction motor. Rotor voltages and rotor currents are not conducted to the rotor but rather are induced into the rotor. The induction motor stator windings produce a magnetic field that rotates at synchronous speed.

The rotational speed of the stator magnetic field depends upon the source frequency and the number of stator poles. The following formula may be used to determine the speed (RPM) of the rotating stator magnetic field:

$$Synchronous\ Speed = \frac{120\ x\ Frequency}{\#\ Poles}$$

Each of the three phases will in turn produce a varying magnetic field and because each of the phases are mechanically and electrically displaced from each other a shifting or rotating of the magnetic field occurs.

Note: The squirrel cage rotor receives its power through transformer action. The magnetic field created by the current flowing in the stator cuts the conductors in the rotor and causes current to flow. The induced current in the rotor circuit will now create its own magnetic field. Motor motion is the result of the interaction between stator and rotor magnetic fields and is based on the principle that like poles repel and unlike poles attract.

AC Polyphase Motors (cont'd)

The speed of the rotor must always be less than the synchronous speed of the stator. If the rotor did turn at the same speed as the stators magnetic field, then the induced current would be zero, and no magnetic flux or torque would be produced. Therefore the rotor must exhibit some "slip" in order to produce torque.

The torque that is developed by an induction motor depends primarily on three factors:

1. Rotor flux (Φ_R)

2. Strength of the stator magnetic flux (Φ_S)

3. Phase angle between the stator flux and the rotor flux (ie. power factor of the rotor - PF_R)

$$T \approx \Phi_R \times \Phi_S \times PF_R$$

Note: Maximum torque occurs when the phase relationship between the rotor current and the stator flux is 45 electrical degrees (power factor is 0.707) and occurs when the rotor's resistance is equal to the rotor's reactance.

The starting and operating torques of squirrel cage motors may be varied by using different rotor construction designs as shown in illustration #178. See table #54 for rotor design classes and their operating characteristics.

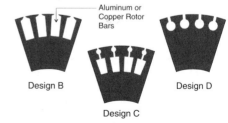

Aluminum or Copper Rotor Bars

Design B

Design C

Design D

Illustration #178 - Rotor Design Classes

AC Polyphase Motors (cont'd)

The power of an induction motor is related to its speed and its torque. Power is the rate (speed) at which work (torque) is being done. The following formulas may be used to determine the output power of a motor:

$$P = T \times S/7.04 \quad \text{(Imperial)}$$
$$HP = T \times S/5252 \quad \text{(Imperial)}$$
$$P = T \times S \quad \text{(Metric)}$$

where:

P = output power expressed in watts
HP = output power expressed in horsepower
T (Imperial) = torque in pound-feet
S (Imperial) = speed in RPM
T (Metric) = torque in Newton-metres
S (Metric) = speed in radians/second

Note: The torque in the power formula refers to the output torque or the torque on the shaft of the motor, and the speed in the power formula refers to the rotor speed.

Wound Rotor Motors

A wound rotor motor differs from a squirrel cage induction motor primarily in the design of its rotor. In a wound rotor motor the rotor consists of insulated coils that are set in slots and connected in a wye arrangement. The rotor circuit is completed through a set of slip-rings, carbon brushes and a wye connected rheostat (adjusting the rheostat will change starting torque and running speeds). Illustration #179 shows a cut-a-way view.

The ability to vary the starting torque of an induction motor by inserting resistance into the rotor circuit allows for smooth and effective starting of high inertia loads. Limited speed control and reduced starting currents are two further advantages of using a wound rotor motor. Adding resistance to the rotor circuit when the motor is running will cause rotor currents to decrease. This in turn will cause the motor to slow down (the rotor flux has decreased, and therefore the rotor torque has decreased).

Wound Rotor Motors (cont'd)

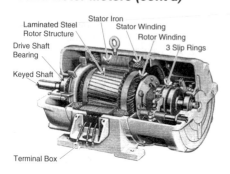

Laminated Steel Rotor Structure
Stator Iron
Stator Winding
Rotor Winding
Drive Shaft Bearing
3 Slip Rings
Keyed Shaft
Terminal Box

Illustration #179 - Wound Rotor Motor

When the rotor's speed is reduced, more voltage is induced into the rotor's winding due to the increased slip (slip = 1 minus the ratio of rotor speed to stator speed).

This in turn will increase current flow in the rotors windings and thereby develop the necessary torque for the motor to run at the lower speed. How great a reduction in speed occurs depends on the characteristics of the load. As much as 50% reduction in speed is possible without the motor becoming unstable or stalling. The lower speed will result in less cooling and therefore higher operating temperatures with reduced efficiency and power but with an increased torque.

Synchronous Motors

Synchronous motors operate at synchronous speed; which means there is no difference in the speed of the rotor and the speed of the rotating magnetic stator field. Synchronous speed is attained by supplying a separate DC voltage to the rotor circuit and producing magnetic poles on the rotor that will lock in on the sweeping stator magnetic poles. See illustration #180.

Synchronous Motors (cont'd)

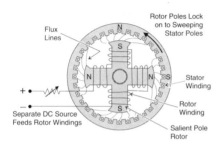

Illustration #180 - Locking Fields on a Synchronous Motor

Adding a load to the motor will cause the rotor to lag the stator by a few electrical degrees producing what is known as the torque angle. Adding more and more load will cause the rotor to lag by a greater amount, until at about 60 degrees the rotor will slip out of synchronism and the motor will stall.

Field voltage failure relays, field current failure relays, automatic resynchronization circuits and out of step relays are some of the synchronizing controls that are required and normally found in synchronous motor controllers. There are two types of rotors used for synchronous motors (see illustration #181A,B):

1. Cylindrical rotors for high speed machines (usually 2 or 4 poles).

2. Salient pole rotors for slow speed machines (usually 8 or more poles).

A synchronous motor requires assistance when starting. The most common starting method involves using a separate squirrel cage winding. This winding may be referred to as a pole-face winding, an amortisseur winding or a damper winding. The winding is shown in illustration #181B. The motor is first started as an induction motor. As the rotor nears its synchronous speed, a DC voltage source is applied to the field windings of the rotor.

Synchronous Motors (cont'd)

High Speed Design (over 500 RPM)

Illustration #181A - Cylindrical Rotor

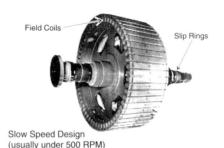

Slow Speed Design
(usually under 500 RPM)

Illustration #181B - Salient Pole Rotor

The magnetic field created by the DC current locks on to the rotating magnetic field of the stator which then brings the rotor up to synchronous speed.

The source of DC for the rotor has in earlier models been a separate DC voltage source connected to the rotor through slip rings and brushes as shown in illustration #181A.

Modern synchronous motors use a brushless excitation system and are therefore well suited for installation in refineries and chemical process plants which require low maintenance and the elimination of sparking in order to meet hazardous location requirements. Illustration #182 highlights a brushless excitation system used on synchronous motors.

Synchronous Motors (cont'd)

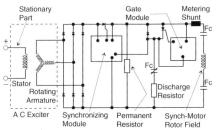

Illustration #182A - Brushless Excitation

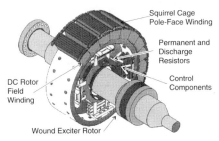

Illustration #182B - Brushless Salient Pole Rotor

The brushless excitation system in illustration #182A consists of a field discharge resistor, diodes, SCR's, a gate module and a synchronizing module. The discharge resistor is shunted across the rotor field during starting. At the proper slip and phase angle, the resistor is removed automatically from the rotor field circuit and DC voltage is applied to the rotor field to pull the motor into synchronism. The switching is done with a field contactor (Fc). Illustration #182B reveals the construction of the brushless rotor and notes key components that are mounted on its assembly.

The nominal voltage rating of a synchronous motor field winding is either 125 or 250 V (DC). Synchronous motors are commonly available in voltages from 240 to 7200 V, (AC) and in horsepower ratings from 40 to 10,000 HP. Synchronous motors are used on continuous load applications such as motor driven generators, compressors, mills, pumps, blowers, dryers, mixers and shredders.

Synchronous Motors (cont'd)

Synchronous motors have the capability to improve the power factor of an electrical distribution system. This may be accomplished by increasing the amount of DC current flowing in the rotor windings beyond that required to operate the motor at synchronous speed.

The advantages of a synchronous motor over a standard induction motor are:

- Generally, lower cost when the rating exceeds one horsepower per RPM
- Adaptability to low speed applications
- Constant synchronous speed operations
- Generally, lower inrush current
- Higher efficiency
- System power factor improvement due to operating at either a 1.0 or 0.8 leading power factor (PF)

Motor Wiring and Connections

Three phase motor windings are connected either in a wye or delta configuration.

Nine leads, as shown in illustration #183, are generally available for wiring a motor to the source (NEMA/EEMAC standard numbering system is shown).

For wye connected motors the ends of each of the individual phases are joined at a common point. In a delta wired motor, the end of each phase is connected to the beginning of the next phase.

By using an ohmmeter it is possible to determine a motors wiring arrangement.

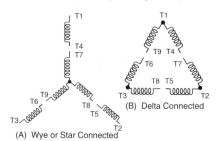

Illustration #183 - Wye and Delta Connections

Motor Wiring & Connections (cont'd)

Continuity exists between three pairs of wires and one set of three wires for wye connected motors.

Continuity exists between three groups of three wires for the delta connected motors. Illustration #184 displays the connections of dual winding wye and delta connected motors.

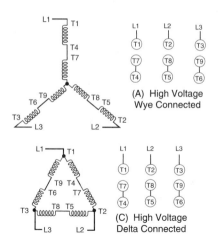

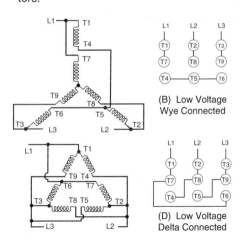

Illustration #184 - Polyphase Motor Connections

Polyphase Motor Speed Control

Speed control of induction motors (not including synchronous motors) is possible by using any one of the following methods:

1. Changing the impressed voltage - The slip and torque of an induction motor will vary by the square of the change in the impressed voltage, as noted in table #56. Reducing the impressed voltage will also reduce the motor's value of breakdown torque and the motor may stall prematurely.

2. Changing the number of poles for which the machine is wound - The speed of an induction motor is inversely proportional to the number of poles. Increasing the number of poles will proportionally decrease the speed. Pole changing may be accomplished by using two or more separate primary windings each having a different number of poles (as shown in illustration #185A) or by reconnecting the windings of specially designed machines called consequent pole motors (as shown in illustrations #185B,C,D).

Consequent pole motors are manufactured with either one or two sets of windings with each winding set providing two different speeds. For example, a motor with a single winding set is connected for either 4 pole or 8 pole operation, producing synchronous speeds of 1800 or 900 RPM. This type of speed control does not lend itself to variable stepless speed control but rather only to fixed values (ie: a cooling fan at low or high speed).

Consequent pole multispeed motors are also designed to provide different operating characteristics. Some are designed to give the same maximum horsepower at all speeds (*CONSTANT HORSEPOWER* motors, the torque will vary inversely with the speed); others are designed to give the same maximum torque at all speeds (*CONSTANT TORQUE* motors); and there are other motors designed to give an increase in torque with an increase in speed (*VARIABLE TORQUE* motors). Illustrations #185B,C,D show the wiring connections for three types of consequent pole motors.

Polyphase Motor Speed Control (cont'd)

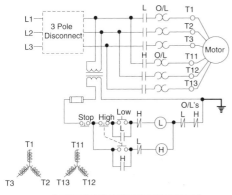

Speed	L1	L2	L3	OPEN
Low	T1	T2	T3	T11, T12, T13
High	T11	T12	T13	T1, T2, T3

(A) 2 Speed, Two Separate Windings

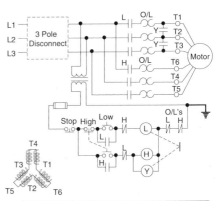

Speed	L1	L2	L3	OPEN	TOGETHER
Low	T1	T2	T3	T4, T5, T6	–
High	T6	T4	T5	–	T1,T2,T3

(B) Constant Torque, 2 Speed, One Winding
(consequent pole)

Illustration #185A,B - Pole Changing Connections

Polyphase Motor Speed Control (cont'd)

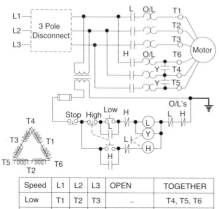

Speed	L1	L2	L3	OPEN	TOGETHER
Low	T1	T2	T3	–	T4, T5, T6
High	T6	T4	T5	T1, T2, T3	

2 Speed, One Winding (consequent pole)

(C) Constant Horsepower

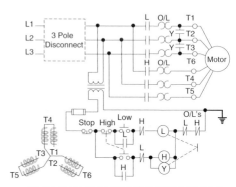

Speed	L1	L2	L3	OPEN	TOGETHER
Low	T1	T2	T3	T4, T5, T6	–
High	T6	T4	T5	–	T1,T2,T3

2 Speed, One Winding (consequent pole)

(D) Variable Torque

Illustration #185C,D - Pole Changing Connections

Polyphase Motor Speed Control (cont'd)

PAM motors (Pole Amplitude Modulation) are not common in North America. Their method of operation is based upon selective switching of coil polarities using a switching arrangement similar to that used for consequent pole machines. Pole ratios other than 2:1 are achieved thereby allowing for intermediate speeds. Externally the motors are connected and switched exactly as consequent pole motors thereby simplifying a retrofit. Unlike a conventional stator, the PAM stator uses irregular coil groupings that produce space harmonics in the rotating field. Space harmonics may cause low starting torque, excessive noise, and sharp dips in torque acceleration; resulting in greater consideration being given to an appropriate winding arrangement. PAM motors are usually more efficient than two winding motors of the same power and speed ratios.

They are also smaller, lighter in weight and generally cheaper than two winding two speed motors. PAM motors also use thin laminations of steel and copper windings as efficiently as normal single speed motors.

3. Changing the frequency of the supply voltage - The speed of an AC motor is directly proportional to the frequency of the power supply. Table #57 reflects the changes in the operating characteristics of AC motors when the frequency changes. When the frequency of the power supply is reduced, the motor will operate at a lower speed, but stator current will increase due to to a lower inductive reactance ($X_L = 2\pi fL$).

The higher stator currents will result in higher copper losses, and therefore the efficiency of the motor will decrease as the frequency decreases. However, if the voltage and the frequency are changed together, then the efficiency of the motor will not decrease. A constant V/Hz ratio results in good operating characteristics and stepless speed control.

Polyphase Motor Speed Control (cont'd)

In the past this constant V/Hz ratio required expensive multimachine variable frequency control equipment. Modern electronic variable-speed motor drives can produce frequencies from 2 to 90 Hz or more from a 60 Hz supply line. Therefore, a four pole motor rated at 1725 RPM can be made to run at speeds from about 60 RPM (2 Hz) to about 2700 RPM (90 Hz).

4. Changing the resistance of the rotor circuit - Inserting resistance in the rotor of an induction motor will decrease the speed of the motor. Decrease in speed is dependent upon the load on the motor. Little or no change in speed is possible for lightly loaded or unloaded motors. This method of speed control is applicable only to wound rotor motors (see wound rotor motor section for more information).

5. Introducing a foreign voltage into the secondary circuit (rotor) - If a foreign voltage is introduced into the secondary circuit of an induction motor, the speed of the motor will change. If the foreign voltage acts in a direction that is opposite to that of the voltage induced in the rotor circuit, the speed of the motor will be reduced; and conversely if the foreign voltage acts in the same direction as the voltage induced in the rotor circuit, the speed will be increased. Additional revolving machines are required to obtain speed control using this method, and therefore this method is used only in special cases as in the speed control of large motors.

Reversing Rotation

The direction of rotation of an induction motor is dependent upon the direction of rotation of the stator flux, which in turn is dependent upon the phase sequence of the applied voltage.

Reversing Rotation (con't)

To reverse the rotation of a three phase motor, interchange any two of the three line leads to the motor, as shown in illustration #186.

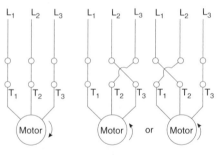

Forward   Reverse

Illustration #186 - Reversing Rotation

Motor Efficiency

Due to the rising costs of electrical energy, considerable emphasis is being placed on the efficiency of electric motors. Motors consume approximately 50% of the electrical energy produced in North America, and any improvement in the efficiency of motors will be of benefit to all. Efficiency is a unit of measure that evaluates a motor's ability to convert electrical energy into mechanical energy at rated voltage and at rated frequency.

The efficiency of an induction motor is equal to the ratio of the useful power out versus the total power into the motor. High efficiency motors require 3 to 10% less electricity to do the same amount of work as standard motors. They are also generally more reliable, last longer and provide lower transformer loading. High efficiency motors are designed with the following enhancements to improve their efficiency:

1. Longer stator and rotor cores that reduce magnetic flux density (saturation).

Motor Efficiency (cont'd)

2. More (larger conductors therefore lower resistance) copper wire in the stator and aluminum in the rotor which improves the motor's conductivity.

3. Air gaps are reduced thus lowering stator currents and stray load losses (magnetic fields that wander).

4. Special alloy steels are used to reduce hysteresis losses, and special laminates are used to reduce eddy currents.

5. Improved stator winding distribution to maximize efficient energy use by reducing or eliminating certain harmonics and thereby reducing core losses.

6. More steel is used in the stator thereby allowing a greater amount of heat transfer out of the motor.

Note: In general high efficiency motors also operate at a higher power factor. See illustration #187.

Power Factor

Power factor is the ratio of active (true) power expressed in watts to apparent power expressed in volt amperes. The closer the power factor is to 1 the lower the line current of the motor. The benefits of operating a motor at .85 or better include lower line voltage drops, lower line power losses, lower kVA demand charges, increased capacity of the distribution system, and lower motor starting disturbances. The costs involved to improve the power factor may be outweighed by the benefits.

The power factor of induction motors will vary depending on their horsepower rating, their frame type (U-frame, T-frame etc.) and if they are of the high efficiency design type. Illustration #187 indicates power factor versus motor horsepower for U and T frame motors.

Power Factor (cont'd)

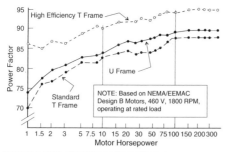

Illustration #187 - U & T Frame Power Factors

Options for improving the overall power factor of an electrical system are to use high power factor motors, install power factor correcting capacitors or add a synchronous motor(s) to the system. Adding capacitors is generally the most economical way to improve the plant power factor.

As motors make up approximately 80% of an industrial plant's load, connecting capacitors into the motor circuit may be the best way to achieve optimum results.

Motors require magnetizing power to operate. If capacitors are installed into the motor circuit, the motor's magnetizing power will be supplied partially or entirely by the capacitors. Where to locate the capacitors in the motor circuit will depend on what type of motor control is being used and the practicality of the location.

Note: Power factor improvement values should not exceed maximum safe values. Excessive improvement may result in high transient voltages, currents and torques which could damage motors and their associated equipment.

Illustration #188 provides four possible locations for the placing of capacitors.

Power Factor (cont'd)

The following motor-capacitor applications should be avoided:

1. Motors that are subject to reversing, plugging, or frequent switching duty. If capacitors are used, locate them as per illustration #188D.
2. Motors that are restarted while still running and generating a back emf.
3. Capacitors that are used with crane or elevator motors where the load may drive the motor, or on multispeed motors.
4. Open transition reduced votage starters that are used with wye-deta connections and capacitors. The capacitors shoud be ocated as shown in iustration #188C.
5. Variable frequency drives do not tolerate power factor correcting capacitors.

There are a number of methods that can be used to determine how many capacitors are required to improve the power factor. One method involves the use of table #61 and the following formula:

kvar of capacitors = coefficient from table #61 x kilowatt input of motor

Note: The kilowatt input can be determined by multiplying the horsepower of the motor by 0.746 and then dividing by the motor efficiency.

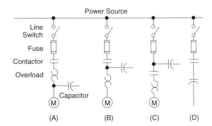

Illustration #188 - Capacitor Locations

Changing the Power Factor										
Original PF (%)	Desired Power Factor (%)									
	80	82	84	86	88	90	92	94	96	98	100
50	0.982	1.034	1.086	1.139	1.192	1.248	1.306	1.369	1.442	1.529	1.732
52	0.893	0.945	0.997	1.050	1.103	1.159	1.217	1.280	1.351	1.440	1.643
54	0.809	0.861	0.913	0.966	1.019	1.075	1.133	1.196	1.267	1.356	1.559
56	0.730	0.782	0.834	0.887	0.940	0.996	1.051	1.117	1.189	1.277	1.480
58	0.655	0.707	0.759	0.812	0.865	0.921	0.976	1.042	1.114	1.202	1.405
60	0.584	0.636	0.688	0.741	0.794	0.850	0.905	0.971	1.043	1.131	1.334
62	0.515	0.567	0.619	0.672	0.725	0.781	0.836	0.902	0.974	1.062	1.265
64	0.450	0.502	0.554	0.607	0.660	0.715	0.771	0.837	0.909	0.997	1.200
66	0.388	0.440	0.492	0.545	0.598	0.654	0.709	0.775	0.847	0.935	1.138
68	0.329	0.381	0.433	0.486	0.539	0.595	0.650	0.716	0.788	0.876	1.079
70	0.270	0.322	0.374	0.427	0.480	0.536	0.591	0.657	0.729	0.811	1.020
72	0.213	0.265	0.317	0.370	0.423	0.479	0.534	0.600	0.672	0.754	0.963
74	0.159	0.211	0.263	0.316	0.369	0.425	0.480	0.546	0.618	0.700	0.909
76	0.105	0.157	0.209	0.262	0.315	0.371	0.426	0.492	0.564	0.652	0.855
78	0.053	0.105	0.157	0.210	0.263	0.319	0.374	0.440	0.512	0.594	0.803
80		0.052	0.104	0.157	0.210	0.266	0.321	0.387	0.459	0.541	0.750
82			0.052	0.105	0.158	0.214	0.269	0.335	0.407	0.489	0.698
84				0.053	0.106	0.162	0.217	0.283	0.355	0.437	0.645
86					0.053	0.109	0.167	0.230	0.301	0.390	0.593
88						0.056	0.114	0.177	0.248	0.337	0.540
90							0.058	0.121	0.192	0.281	0.484
92								0.063	0.134	0.223	0.426
94									0.071	0.160	0.363
96										0.089	0.292
98											0.203

Table #61 - Coefficient for Power Factor Correction

Power Factor (cont'd)
Power factor correction example:

A 100 HP motor operating at a power factor of 0.76 (76%) has a full load efficiency of 80%. Calculate, using table #61, the size of capacitor required to improve the power factor to 0.9 (90%):

100 HP = 100 x 0.746 ... = 74.6 kW

Kilowatt input = 74.6/.8 ... = 93.25 kW

From table #61 the coefficient is 0.371

kvar of capacitor = 0.371 x 93.25 ... = 34.6

Low Voltage Polyphase Motor Control

For every AC motor there is a motor starter nearby that controls and protects the motor and the motor circuit. NEC article 430 and CEC Section 28 detail the minimum code requirements for motors, motor controllers, motor circuit conductors, and motor circuit protection. There are three principal ways of starting polyphase motors:

1. Full voltage (across the line)
2. Reduced voltage (electromechanical):
   a. autotransformer (most common)
   b. part winding
   c. wye-delta
   d. resistance (not commonly used)
   e. reactor (not commonly used)
3. Solid-state Controller - provides both reduced starting voltage and stepless speed control

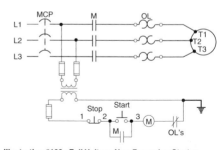

Illustration #189 - Full Voltage Non-Reversing Starter

LV Polyphase Motor Control (cont'd)

Full voltage non-reversing electromagnetic starters are the most commonly used controllers for both polyphase and single phase motors. As shown in illustration #189, the controller and its associated equipment provide both on-off control and protection. The overcurrent device is a molded case circuit breaker called a motor circuit protector (MCP) which provides protection against short circuits only.

The starting current of a motor is many times greater than its operating full load current, and overcurrent devices are sized to avoid nuisance tripping. Overload protection is provided by sensing the operating current and causing the control circuit to be interrupted if a sustained overload should occur.

The overload relays normally require a manual reset after an overload trip condition.

Overload devices are classified and described as follows:

1. Thermal:

a. Eutectic alloy: A spring loaded mechanism that is held rigid by a eutectic alloy substance. If the current generates enough heat the alloy melts, thereby releasing the mechanism and opening the control circuit.

b. Bimetal: A bimetal strip will deflect from the normally closed contact position if the load current increases and produces enough heat.

c. Time delay fuse: Overload current will open a fuse link that has an inverse time operating characteristic.

2. Magnetic:
A magnetic overload relay has a movable magnetic core inside a coil which carries the motor current. The current passing through the coil will cause the core to move upwards and if it rises far enough a set of contacts will trip, opening the control circuit.

LV Polyphase Motor Control (cont'd)

3. Solid state: Modern starters are making greater use of current transformers to sense the motor current in each of the lines and then to act upon the values measured. Solid state current sensors used in conjunction with electronic processors provide not only overload protection but also phase unbalance, ground fault and remote circuit monitoring.

Overload heaters are no longer required, and overload settings are conveniently set by the adjustment of dip switches, as shown in illustration #190.

NEMA/EEMAC publish standard ratings for motor controllers as shown in table #62 (non-plugging and non-jogging duty) and table #63 (plugging and jogging duty).

IEC rated controllers are now becoming more common in North America due to their low cost, small size and light weight. IEC standards specify the types of tests that the manufacturer must pass in order to obtain IEC ratings. This is in contrast to NEMA/EEMAC requirements where standard rating tables (tables #62 and #63) are used to select controllers.

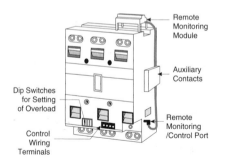

Remote Monitoring Module

Auxiliary Contacts

Dip Switches for Setting of Overload

Remote Monitoring /Control Port

Control Wiring Terminals

Illustration #190 - Electronic Current Sensing Controller

LV Polyphase Motor Control (cont'd)

Three Phase, single speed, full voltage, magnetic controllers for non-plugging , non-jogging duty (Horsepower)					
Controller size	Continuous current rating (Amps)	60 Hertz		50 Hertz	60 Hertz
		200 V	230 V	380 V	460 or 575 V
00	9	1 1/2	1 1/2	1 1/2	2
0	18	3	3	5	5
1	27	7 1/2	7 1/2	10	10
2	45	10	15	25	25
3	90	25	30	50	50
4	135	40	50	75	100
5	270	75	100	150	200
6	540	150	200	300	400
7	810		300		600
8	1215		450		900
9	2250		800		1600

Table #62 - NEMA/EEMAC Motor Starters

LV Polyphase Motor Control (cont'd)

Three Phase, single speed, full voltage, magnetic controllers for plug-stop, plug-reverse or jogging duty (Horsepower)					
Controller size	Continuous current rating (Amps)	60 Hertz		50 Hertz	60 Hertz
		200 V	230 V	380 V	460 or 575 V
0	18	1 1/2	1 1/2	1 1/2	2
1	27	3	3	5	5
2	45	7 1/2	10	15	15
3	90	15	20	30	30
4	135	25	30	50	60
5	270	60	75	125	150
6	540	125	150	250	300

Table #63 - NEMA/EEMAC Motor Starters

To choose an IEC rated controller, the user must first identify the utilization category (see table #64) based on its duty and then choose a controller that has the capability to handle the intended load and duty.

The user must then check that the manufacturer's listed contact life is adequate and appropriate for the load and duty.

LV Polyphase Motor Control (cont'd)

Utilization Categories	
Category	**Typical Duty**
AC1	Non Inductive or slightly inductive loads
AC2	Starting of slip-ring motors
AC3	Starting of squirrel cage motors and switching off only after motor is up to speed
AC4	Starting of squirrel cage motors with inching and plugging duty

Note: In an AC3 application, the contactor will never interrupt more than the motor's full load current. If the application requires interruption of current greater than the full load current, it is an AC4 application.

Table #64 - IEC Utilization Categories

IEC rated devices are generally rated closer to their ultimate capabilities than NEMA/EEMAC based products, and therefore they are smaller in physical size when compared to NEMA/EEMAC products of similar ratings.

Protecting IEC rated controllers becomes critical because they they do not have the same short circuit withstand capabilities as NEMA/EEMAC rated contactors. To protect these controllers fast acting or current limiting fuses and/or circuit breakers are used.

LV Polyphase Motor Control (cont'd)

Full voltage reversing starters are commonly used to control loads like overhead cranes, lathes, mixers and drill presses, or when a plugging operation is needed for sudden motor braking. A schematic for a reversing motor controller is shown in illustration #191. A mechanical interlock between the forward and reverse contactor protects against accidental closure of both sets of power contacts.

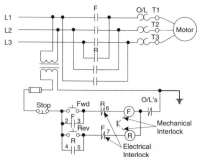

Illustration #191 - Full Voltage Reversing Starter

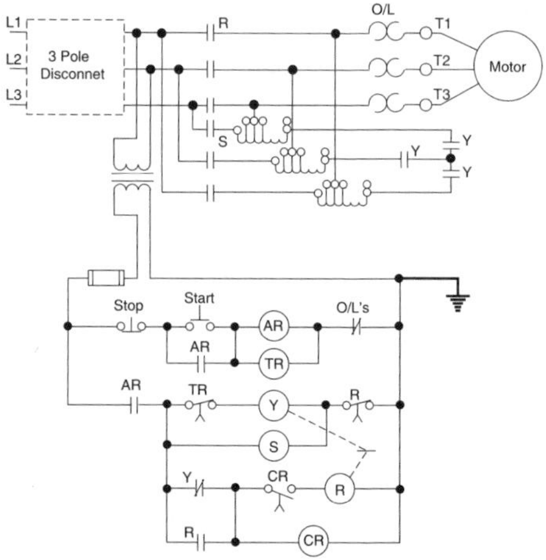

Illustration #192 - Autotransformer Starter

LV Polyphase Motor Control (cont'd)

Reduced voltage motor controllers are primarily used to reduce large inrush currents during starting and/or to dampen starting torques. The autotransformer type of motor controller is by far the most popular electro-magnetic means to reduce starting currents and torques in North America. The part winding controller, wye-delta controller and the solid-state controller are also used.

Typical Characteristics for NEMA Design B Motors								
Starting Method	**% Voltage at motor terminals**	**Motor Starting Current (%)**		**Line Side Current (%)**		**Starting Torque (%)**	
		Locked Rotor	**Full Load**	**Locked Rotor**	**Full Load**	**Locked Rotor**	**Full Load**
Full Voltage	100	100	600	100	600	100	180
Autotrans.							
80% tap	80	80	480	64	384	64	115
65% tap	65	65	390	42	252	42	76
50% tap	50	50	300	25	150	25	45
Part Winding	100	65	390	65	390	45	81
Wye-Delta	100	33	198	33	198	33	60
Solid-State	0-100	0-100	0-600	0-100	0-600	0-100	0-180

Table #65 - Reduced Voltage Starter Characteristics

LV Polyphase Motor Control (cont'd)

Table #65 provides a comparison of the characteristic differences between each of the reduced voltage starters.

Illustrations #192 (autotransformer), #193 (part winding) and #194 (wye-delta) provide schematic diagrams for the controllers.

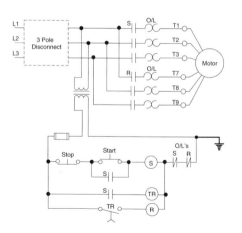

Illustration #193 - Part Winding Starter

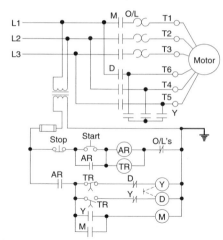

Illustration #194 - Wye-Delta Starter

Solid State Controllers and VFDs

Electronic soft start controllers and variable frequency drives (VFDs) are quickly displacing conventional electromagnetic reduced voltage controllers. Electronic controllers are cost effective and adapt well to communicating electronically with data control systems. VFDs permit motors to drive their loads over a wide speed range thereby optimizing on loading requirements.

Effect of Unbalanced Line Voltages

A motor operating from a three phase unbalanced voltage source will experience an increase in its operating temperature.

Small voltage unbalances may be enough to cause the motor to overheat and fail prematurely.

A 1% voltage unbalance may result in a 6 - 10% current unbalance. The unbalanced currents will produce additional heating in both the stator and rotor windings and shorten the life of the motor insulation.

NEMA/EEMAC have established derating values for motors that must operate at rated load with unbalanced voltages (see illustration #195).

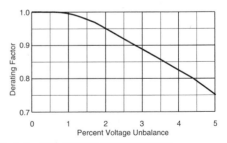

Illustration #195 - Unbalanced Voltage Derating

Unbalanced Line Voltages (cont'd)

The increase in temperature due to unbalanced voltages at rated load is approximately equal to two times the square of the percentage voltage unbalance and can be expressed as follows:

%Change in temperature
$$= 2 \times (\text{\%voltage unbalance})^2$$

Example: The temperature change in percent for a fully loaded motor being supplied with a 3% voltage unbalance is: $2 \times (3)^2 = 18\%$

The worst case of voltage unbalance is a single phase condition. When a single phase condition occurs, the currents in the remaining two lines increases by 1.732 times. If a motor is operating at 40% or less of its rated load when the single phase condition occurs, the motor will probably not overheat. Any single phase loading values above 40% will result in the motor quickly overheating and prematurely failing.

The following list reveals some of the sources and causes of voltage unbalances:

a. Large single phase loads that are connected to a three phase system
b. Poor electrical connections
c. Blown fuses (single phasing)
d. Improperly impedance matched three phase source transformers
e. Open delta connected transformers that are supplying motors
f. Harmonics
g. Electronic motor controller output to the motor is unbalanced

Single Phase Motors

There are three classes belonging to this group of motors:

1. **Induction motors** of which there are the following common types: split phase resistance start, split phase capacitor start, permanent split capacitor motor and shaded pole.

Single Phase Motor (cont'd)

2. *Universal motors*
3. *Hysteresis,* also called hysteresis-syn-chronous motors

Single phase motors are either classified as being fractional horsepower (1/10 HP or less to 3/4 HP) or integral horsepower (1 HP to 10 HP). Nameplate data appearing on single phase motors is comparable to that found on polyphase motors (see illustration #168).

Split-Phase Motors

The split phase resistance start induction motor, unlike the polyphase motor, is not self starting. To approximate the starting characteristics of a polyphase machine, two independent sets of magnetic fluxes that are out of phase in time and space by (ideally) 90 degrees need to be created in the stator circuit of the motor.

To accomplish this starting characteristic, two separate windings connected in parallel are used.

The first winding, called the run or main winding, has a low resistance and a high inductance.

The second winding, called the start or auxiliary winding, has a high resistance and a much lower inductance. When both windings are energized, the current in the start winding is more in phase with the voltage than the run winding.

The result is that the single phase line current is split into two parts with the run winding current lagging behind the start winding current. This produces a shifting or rotating magnetic field. The pulsating rotating stator field induces a voltage into the rotor, causing a rotor current to flow resulting in rotor flux.

The rotor flux tries to lock on to the shifting or rotating stator flux thereby developing motor torque. When the rotor achieves approximately 75% of its rated speed, the torque developed by the run winding alone is equal to the torque of both the start and run winding acting together.

Split-Phase Motors (cont'd)

At rated speed the torque developed by both windings in the circuit is less than when only the run winding is energized. As a result, a centrifugal switch or an electromagnetic current relay (see illustration #196) is used to remove the start winding from the motor circuit when 75% of the rated speed of the motor has been reached. The direction of rotation of a split phase motor is dependent upon how the run and start windings are connected to the source and to each other.

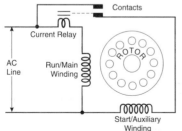

Illustration #196 - Electromagnetic Current Relay

The power factor of the start winding is higher than that of the run winding; therefore, the start winding flux will always lead the run winding flux. This means that the field will revolve in a direction from a given start winding pole to an adjacent run winding pole of the same polarity (to reverse rotation simply reverse the start winding).

The starting torque of a split phase induction motor can be represented by an equation:

$$T_{START} = k \times I_{RW} \times I_{SW} \times \sine \theta$$

k = machine constant
I_{RW} = current in run winding
I_{SW} = current in start winding
θ = phase displacement in degrees
between I_{RW} and I_{SW}

Capacitor Split Phase Motor

A capacitor split phase motor develops a much larger starting torque than a resistance start motor due to the presence of a larger phase displacement in electrical degrees between the run and start windings.

Split-Phase Motors (cont'd)

The phase angle between the start and run winding varies between 75 and 88 electrical degrees for a capacitor start motor as compared to 25 and 30 degrees for a resistance start split phase motor.

In the capacitor split phase motor, a starting capacitor is connected in series with the start winding as shown in illustration #197.

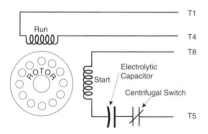

Illustration #197 - Capacitor Start Motor

A capacitor start motor has high starting torque (300% of rated) and good speed regulation. These qualities make the capacitor start motor well suited for coal stokers, compressors, and reciprocating pumps; whereas, the resistance start split phase motor lends itself well for washing machines, drill presses, small fans and blowers.

Permanent Split Capacitor Motor

This motor does not have a centrifugal switch; therefore, the start winding and its capacitor are permanently connected to the run winding. The permanent split capacitor motor's operation closely resembles that of a two phase motor. One benefit of this type of operation is that the rotating field set up by the two stator windings is more uniform, and the motor is less noisy under load. In comparison to other split phase motors, the permanent split capacitor start motor has:

a. a higher full load efficiency

Split-Phase Motors (cont'd)

b. a higher power factor at full load
c. a lower full-load line current
d. an increased pull out torque

However, the permanent split capacitor motor does have some disadvantages, such as, low starting and locked rotor torque (75% to 125%) and a higher initial cost. This motor lends itself well to applications for shaft mounted fans used in unit heaters and for ventilating fans.

Motor Lead Numbering & Color Code

For single phase and split phase motors, NEMA/EEMAC have established a standardized numbering and color coding system to assist in the identification of motor leads. The number of each lead is prefixed with either the letter "T" or the letter "P". Illustration #198A depicts the code for a split phase motor having one run winding and one start winding. Illustration #198B shows the same motor with integral thermal protection included.

Illustration #198C provides the NEMA/EEMAC standard for a dual voltage motor with thermal protection. Illustration #198D identifies the NEMA/EEMAC color code that can be used rather than numbers.

Single Voltage Split Phase Wiring

A connection diagram for a single voltage split phase motor is shown in illustration #199A. The connections required to reverse the motor are shown in illustration #199B.

Note: To reverse the rotation of a split phase motor, simply reverse the direction of current flow through the start windings by interchanging the two starting winding leads.

Low/High Voltage Split Phase Wiring

Illustration #200A shows the connections for a split phase motor wired for the lower voltage, and illustration #200B shows the connections for the same split phase motor connected for the higher voltage.

Split-Phase Motors (cont'd)

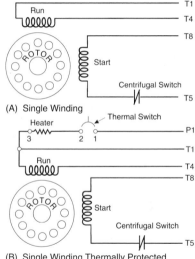

(A) Single Winding

(B) Single Winding Thermally Protected

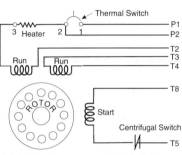

(C) Dual Voltage Thermally Protected

T1 - BLUE	T5 - BLACK
T2 - WHITE	T8 - RED
T3 - ORANGE	P1 - NOT ASSIGNED
T4 - YELLOW	P2 - BROWN

(D) Color Code for Numbered Terminals

**Illustration #198 - Numbering & Color Code for
Single Phase Motors**

Split-Phase Motors (cont'd)

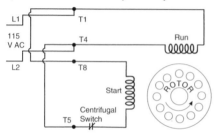

Illustration #199A - Counter Clockwise Rotation

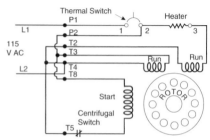

Illustration #200A - Lower Voltage Connections

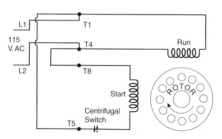

Illustration #199B - Clockwise Rotation

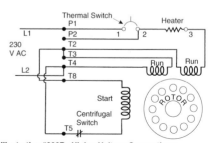

Illustration #200B - Higher Voltage Connections

Speed Control

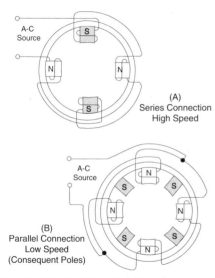

(A)
Series Connection
High Speed

(B)
Parallel Connection
Low Speed
(Consequent Poles)

Illustration #201 - Single Phase Speed Control

Speed control of split phase motors may be accomplished by using either another set of windings or by using the consequent pole concept as shown in illustration #201.

Shaded Pole Motor

The shaded pole motor is an induction motor that utilizes a short-circuited coil or copper ring called a shading coil to produce the starting torque. The shading coil is wound around one end of the stator pole face as shown in illustration #202.

Current flowing in the primary stator windings will produce a magnetic field which induces a voltage into the shading coil and causes a high current to flow in the shaded pole coil.

Shaded Pole Motor (cont'd)

The magnetic field that is developed by the shaded pole current now opposes the magnetic field that initially created it. This shaded coil field is nearly 90 electrical degrees behind the main magnetic field, and the interaction of these two magnetic fields results in rotational torque.

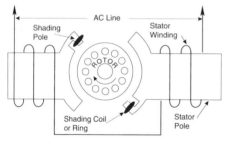

Illustration #202 - Shaded Pole Motor

The direction of rotation is determined by the shading coils with the motor turning toward the shaded pole. Reversing a shaded pole motor is very difficult. It involves either disassembling the motor and reversing the stator by turning it end for end or moving the shading coils.

Shaded pole motors are available with two, four, six or eight poles. The speed of the shaded pole motor may be varied by using a multiple winding motor and using the slip characteristics of the motor to effect speed change. This method of speed control is only effective when the motor is loaded.

Shaded pole motors are typically used to drive small fans, blowers, small kitchen appliances, turn-tables, clocks and valve operators where there is a low horsepower requirement (1/100 to 1.5 HP). Shaded pole motors are simple in construction and do not require auxiliary parts such as capacitors and centrifugal switches.

Shaded Pole Motors (cont'd)

Shaded pole motors are reliable, rugged and inexpensive, but they do have low starting torques, poor efficiencies (5% to 35%) and operate at low power factors.

Universal Motors

A universal motor is a redesigned DC series motor that performs essentially in the same way whether it is connected to an AC 60 Hz source or to DC. The change made to this motor that makes it satisfactory for AC as well as DC operation is in the reduction of the number of turns of wire in the series field plus an increase in the number of coils in the armature circuit.

Because the armature conductors and the field windings are connected in series, as shown in illustration #203, the current through both of them is the same. There will be no time lag or phase displacement in their magnetic fields, and the attraction and repulsion forces will be nearly identical whether operating on AC or on DC.

Compensating windings which reduce armature reaction are almost always used in universal motors.

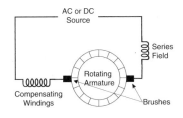

Illustration #203 - Universal Motor

Universal motors develop more horsepower per unit weight than all other single phase motors. Horsepower ratings range from 1/100 HP to 1 HP. They exhibit excellent starting torques and have a high value of stall or breakdown torque.

Universal Motor (cont'd)

Full load operating speeds are usually in the range of 4000 to 16000 RPM and may be over 20,000 RPM. Most universal motors are coupled or gear reduced to their loads. Speed control is easily accomplished in universal motors by varying the armature voltage as shown in illustration #204.

To reverse a universal motor, simply change the direction of current flow through either the series field winding or through the armature winding, but not through both. Reversible motors have four or five leads available for connection and require the use of a reversing switch. Universal motors are used in power tools such as routers, power saws, drills and are commonly found in household appliances such as vacuum cleaners, blenders and food processors.

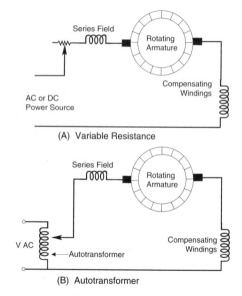

Illustration #204A,B - Universal Motor Speed Control

Universal Motors (cont'd)

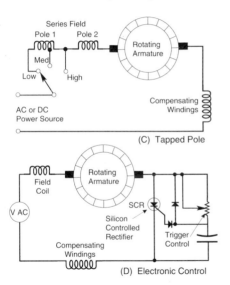

Illustration #204C,D - Universal Motor Speed Control

Hysteresis Motor

The stator of a hysteresis motor (also called a hysteresis-synchronous motor) is the same as that of an induction motor and often is of the shaded pole design. The rotor consists of a smooth cylinder made of a very hard permanent magnet alloy material like cobalt steel (see illustration #205).

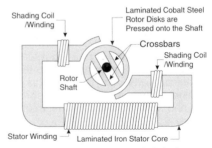

The hysteresis motor starts as a shaded-pole induction motor and runs as a synchronous motor.

Illustration #205 - Hysteresis Motor

Hysteresis Motors (cont'd)

Starting torque is obtained by the creation of a shifting stator flux initiated by the interaction of the stator winding flux and the shading coil flux. A voltage is induced into the rotor resulting in magnetic poles being formed in the rotor. The rotating stator magnetic field exerts a torque on the induced magnetic poles of the rotor, and as the stator poles rotate the rotor follows.

The energy transferred to the rotor by the rotating flux of the stator is in the form of **hysteresis energy (molecular heating due to magnetic retentivity)**. If the rotor is allowed to rotate, this energy is converted into mechanical power. The torque of this motor is constant from standstill to synchronous speed and is independent of either frequency or speed.

Synchronous speed is attained when the rotor poles, which have a high value of magnetic retentivity, lock onto the rotating stator poles. The torque at synchronous speed is no longer constant but varies with the load. Hysteresis motors are very smooth and quiet in their operation. Starting current is approximately 150% of rated full load current.

Because a hysteresis motor may synchronize any load that it can accelerate, and the transition to synchronization is done without abrupt changes, the motor is suitable for use in tape recorders, phonographs and in drives used to manufacture certain products.

Motor Maintenance

There are two types of motor maintenance programs; corrective maintenance which is performed when a problem arises, and preventive maintenance which is performed to avoid problems from arising. A well planned and well run maintenance program will improve overall plant and equipment productivity, minimize downtime and reduce maintenance expenditures. Maintenance of electric motors requires a good knowledge of how the motor works and how the motor should behave when operating normally. Motor failures may be broken down into three groups:

1. Failures associated with mechanical problems.
2. Failures associated with lack of maintenance and general deterioration.
3. Failures associated with the quality of the electrical power supply.

Mechanical Problems

a. ***Motors drive mechanical loads that are either coupled or belted to the motor.*** The motor must not only have the required horsepower for the load but must also have the required torque and acceleration energy to bring the load up to speed. Occasionally the motor and the load are mismatched. If the motor torque is insufficient to properly accelerate the load, then the motor winding will overheat and eventually fail.

b. ***Misalignment of belts or couplings,*** as shown in illustration #206, and excessive vibration from connected piping or equipment may lead to severe overloading of the motor and its early failure. Compensation for changes in alignment due to temperature changes should also be made. All motor bases should be firmly and securely fastened down to prevent shifting and excessive vibration. Vibration can also be caused by unbalanced pulleys or improperly fastened anchor bolts.

Motor Maintenance (cont'd)

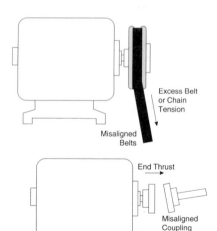

Excess Belt
or Chain
Tension

Misaligned
Belts

End Thrust

Misaligned
Coupling

Illustration #206 - Motor Load Misalignments

c. **Bearings are used in all motors.** On occasion they must be checked and/or lubricated. Bearings that are worn or failing create an extra load for the motor, causing it to work harder in order to drive the load. In addition to the motor bearings, the bearings of the equipment that is being driven by the motor should be checked as well.

d. **The thermal capacity of the motor windings** is a limiting factor on how much load the motor can carry. Providing a means of removing heat created in a motor increases both the capacity and the life of the motor. Ventilation of the motor is important in removing heat, and anything that restricts ventilation will cause the motor to operate at a higher than desired temperature. Sawdust, blown grass and weeds, dirt, dust, and even snow may restrict the ventilation of the motor. Overlubricating a motor and thus coating the windings with oil will also add to the heating problem.

Motor Maintenance (cont'd)

e. Placing a motor in a pit or area with little or no air circulation will reduce ventilation. Higher altitudes may also reduce the degree of ventilation; motor manufacturers should be consulted for derating factors when motors are located at higher altitudes.

Lack of Maintenance/Deterioration

Deterioration at an accelerated rate is often caused by moisture, poor lubrication, and by the presence of insects and small animals.

a. *The insulating quality of a motor* may be reduced if moisture forms on the windings. On lower voltage motors this somewhat reduced value of insulation resistance may not pose much of a problem, but for high voltage motors it is critical that motors operate at maximum insulation resistance values. If these values are not maintained the motor may experience a winding-to-winding short.

Enclosures are built to prevent the entrance of liquids from the outside, however moisture can still form on the inside due to condensation. A motor that is shut down after operating for a period of time will have moisture form on its internal parts due to condensation. Often motors come equipped with space heaters that operate when the motor is shut down to prevent condensation.

b. *Overlubricating is one of the principle causes of motor bearing failures*. Periodic lubrication is necessary and is dependent on the size and duty of the motor as noted in table #66.

c. *Insects and small animals may occupy the space* within a motor and reduce either the ventilation or feed upon the insulation itself. Rodents have been known to knaw or urinate on the insulation thereby destroying its insulating qualities. A simple solution to this problem is to make certain that all screens and covers are kept in place on the motor enclosure.

Motor Maintenance (cont'd)

Motor Lubrication Schedule			
	Horsepower		
Duty	0.5-7.5	10-40	Over 40
Light. Motor operates infrequently (1hr./day)	10 yr	7 yr	5 yr
Standard. Machine tools, fans and pumps are good examples	7 yr	5 yr	3 yr
Severe. Motor operates continuously in locations subject to severe vibrations	4 yr	2 yr	1 yr
Very severe. Motor is subjected to a dirty environment and high vibration. The shaft of the motor overheats	9 mo	8 mo	6 mo

Table #66 - Motor Lubrication Schedule

Electrical Failures

a. *Voltage unbalances and single phasing* in AC polyphase motors may result in both stator and rotor overheating (see illustration #195 and the accompanying text mentioned earlier in this section concerning this problem).

b. *Low voltage applied to a motor* may cause the motor to fail prematurely. When the voltage goes down, the current must go up in order to meet the load demands.

c. *High voltage, more than 10%* above rated nameplate value, will cause the iron in the motor stator to saturate. The magnetically saturated stator now has a lower impedance, and the current will rise causing the motor to overheat. High efficiency "T" frame motors are especially susceptible to the heating effects of overvoltages; overvoltages of more than 5% cannot be tolerated.

Motor Maintenance (cont'd)

d. **Lightning and other surge voltages** can puncture the insulation on motor windings. Surge capacitors are often employed to clamp sudden overvoltages. To make surge capacitors effective for both transient and lightning surges, the connecting leads from the capacitor to the motor terminals should be no more than 12 inches (305 mm) in length.

e. **Excessive starting of the motor** may lead to early motor failure. Each time an AC motor starts, six to ten times its rated current will flow through the stator windings. This current adds to the heating of the motor, and if sufficient time is not allowed for the motor to cool between starts, the motor may overheat and fail. Most motor manufacturers will not guarantee their motors for starting more than 10 times an hour. Large motors are usually limited to two starts per hour.

f. **Loose electrical connections** invite problems for motors. Loose connections result in unwarranted voltage unbalances, voltage drops, motor lead heating, and sporadic motor performance.

Maintenance Scheduling

The frequency and extent of required maintenance depends on the type, service and application of the motor; however, all motors should be periodically checked for the following:

Cleanliness: The cleaner the motor the cooler and more efficient its operation. Periodically all motors should be either wiped clean or disassembled and cleaned. Use a nonconducting detergent when washing motor parts, and thoroughly dry the motor before restoring it into service.

Electrical connections: Connections should be secure, tight, insulated and corrosion free. Clean, retighten and reinsulate all electrical connections.

Maintenance Scheduling (cont'd)

Proper ventilation and acceptable ambient temperatures: There should be no constrictions or blockage of ventilation to the motor, and the ambient temperature in which the motor is found should be within its acceptable operating range. If high ambient temperatures are present, then additional ventilation or air conditioning may be needed. Remove dust, dirt, snow and any other blanketing element from the motor enclosure.

Tightness of mounts and alignment of the motor with the load: Make the necessary adjustments to realign the motor. Check shaft keys for looseness, and look for worn gears, chains and belts.

Proper lubrication and bearing wear of motor and its load: Listen for excessive bearing noise, and follow a regular lubrication schedule (see table #66).

Note: Do not over lubricate

Check the coloration and wear of the motor windings: Darkened sections of the motor winding may indicate overheating due to either overload, single phasing, overvoltages, locked rotor or unbalanced voltages.

Check the condition of the field windings on DC motors: Serious overspeeds will result if the field winding open circuits. Check the connections and condition of the field winding.

Check the condition of auxiliary mechanical parts: Slip rings, commutators and brushes all experience wear and should be checked periodically for wearing and contaminants.

Check the condition of the rotating assembly: Rotors and rotating armatures should be inspected periodically. Check for overheating of the rotor bars or armature conductors. Check for alignment and clearances between the rotating structure and the stationary structure (air gap).

Maintenance Scheduling (cont'd)

Check switches: Centrifugal switches are usually the first devices to fail in single phase motors and should be checked and cleaned periodically.

Check auxiliary motor equipment: Starting or running capacitors and power factor correction capacitors should be checked for leakage or swelling.

Note: Many motor operational checks may be done quickly and effectively by using the senses of sight, touch, hearing and smell. Common sense coupled with experience are valuable tools in predicting and evaluating the state or condition of a motor.

Troubleshooting

Troubleshooting involves a process that starts by recognizing the major components in a motor system. A motor system consists of four major components:

- the power supply
- the controller
- the motor
- the motor load

When a problem occurs, it is necessary to first determine in which part of the system the problem lies. Tables #67, #68, #69, #70 and #71 may serve as a guide to determine some of the probable causes of a motor circuit problem.

Synchronous Motor Troubleshooting

Troubleshooting - Synchronous Motor Problems	
Symptom - Probable Cause	**Symptom - Probable Cause**
Fails to start	**Runs slow**
Blown fuses or tripped circuit breaker	Low frequency
Open in one phase	**Pulls out of synchronism**
Overload	Overload
Low voltage	Open field coils
Runs hot	No exciter voltage
Overload	Open field in rheostat
Clogged ventilating ducts	Rheostat resistance set too high
Short circuited stator coils	**Will not synchronize**
Open stator coils	Field current set too low
Overvoltage	Open in field coils
Grounded stator	No exciter voltage
Field current set too low	Open in field rheostat
Field current set too high	**Vibrates severely**
Uneven air gap	Out of synchronism
Rotor rubbing on stator	Open armature coil
Runs fast	Open phase (single phasing)
High frequency	Misaligned to load

Table #67 - Troubleshooting Synchronous Motors

Polyphase Induction Motor Troubleshooting

Troubleshooting Polyphase Squirrel Cage Induction Motor Problems	
Symptom - Probable Cause	**Symptom - Probable Cause**
Fails to start	One phase open
Blown fuses or tripped circuit breaker	Grounded stator
Open in one phase	Uneven air gap
Overload	Rotor rubbing on stator
Runs hot	Voltage unbalances
Overload	**Runs slow**
Clogged ventilating ducts	Overload
Inadequate ventilation	Undervoltage
Short circuited stator coils	Low frequency
Undervoltage	Broken rotor bars
Overvoltage	Short circuited stator coils
Low frequency	Open stator coils
Open stator coils	One phase open (single phasing)

Table #68 - Troubleshooting Polyphase Squirrel Cage Induction Motors

Wound Rotor Induction Motor Troubleshooting

Troubleshooting - Wound Rotor Induction Motor Problems	
Symptom - Probable Cause	**Symptom - Probable Cause**
Fails to start	Open stator coils
Blown fuses or tripped circuit breaker	One phase open
Open in phase of stator	Low frequency
Overload	Grounded stator
Open in secondary controller circuit	Rotor rubbing on stator
Inadequate brush tension	Voltage unbalances
Brushes do not touch collector rings	**Runs slow**
Open rotor circuit	Overload
Runs hot	Undervoltage
Overload	Low frequency
Clogged ventilating ducts	Too much resistance in secondary controller circuit
Undervoltage	Short circuited stator coils
Overvoltage	Open stator coils
Uneven air gap	One phase open (single phasing)
Short circuited stator coils	Open in rotor circuit

Table #69 - Troubleshooting Wound Rotor Induction Motors

Single Phase Motor Troubleshooting

Troubleshooting - Single Phase Motor Problems	
Symptom - Probable Cause	**Symptom - Probable Cause**
Fails to start	Overvoltage
Blown fuses or tripped circuit breaker	Clogged ventilating ducts
Defective starting mechanism	Short circuited stator coils
Open in start winding	Worn bearings
Open in run winding	Low frequency
Short circuited capacitor	Rotor rubbing on stator
Open capacitor	**Runs slow**
Overload	Overload
Seized bearing	Undervoltage
Runs hot	Low frequency
Overload	Broken rotor bars
Starting mechanism does not open	Short circuited stator coils
Undervoltage	

Table #70 - Troubleshooting Single Phase Motors

DC Motor Troubleshooting

Troubleshooting - DC Motor Problems	
Symptom - Probable Cause	**Symptom - Probable Cause**
Fails to start	Open armature coil
Blown fuse or tripped circuit breaker	Commutator not geometrically centered
Frozen shaft or seized bearing	High mica or high commutator bars
Brushes not in contact with commutator	Vibration
Open in shunt field	Short circuited or reversed commutator
Open in armature circuit	winding
	Runs fast
Runs hot	Overvoltage
Overload	Series field bucking shunt field
Short circuited armature	Open in shunt field of a lightly loaded
Clogged ventilating ducts	compound motor
Brushes off magnetic neutral plane	Shunt field rheostat resistance set too
Overvoltage or undervoltage	high
Short circuited field coils	Shunt field coil short circuited
High ambient temperature	**Runs slow**
	Undervoltage
Sparks at brushes	Overload
Commutator and/or brushes are dirty	Short circuit in armature
Wrong brush grade	Brushes off magnetic neutral plane
Brushes off magnetic neutral plane	Starting resistance of starter not cut out
Wrong brush adjustment	

Table #71 - Troubleshooting DC Motors

SECTION
EIGHT
TRANSFORMERS

Introduction

Transformers are an integral and invaluable part of AC electrical distribution systems. Transformers serve a key role in providing voltages and currents to meet customer needs and demands.

Electrical energy is usually generated at a voltage value that does not meet utilization requirements. Therefore, numerous transformers are needed between the alternator and the load. Each transformer serves a vital role in delivering electrical energy. In illustration #207 voltages are increased or decreased as needed by using transformers.

High transmission voltages (115 kV and up to 1000 kV) are used when transporting electricity over great distances. These high voltages are then transformed to lower voltages at or near the load.

Operating Principles

A transformer is a static electromagnetic machine with no moving parts.

A transformer transfers electrical energy from one circuit to another by using the principle of induction. In its most simple form, the transformer consists of two sets of windings wound on one common laminated iron core, as shown in illustration #208. The winding that is connected to the source voltage is known as the *primary winding*, and the winding that is connected to the load is known as the *secondary winding*. Electrical energy is transferred from the primary winding to the secondary winding by magnetic induction. See Section One for more information on induction.

An alternating primary voltage causes an alternating primary current to flow, producing a constantly changing primary magnetic field. This primary magnetic field cuts through the secondary windings and creates a voltage across the secondary. The secondary circuit of a transformer reflects the values of the primary or source side.

Operating Principles (cont'd)

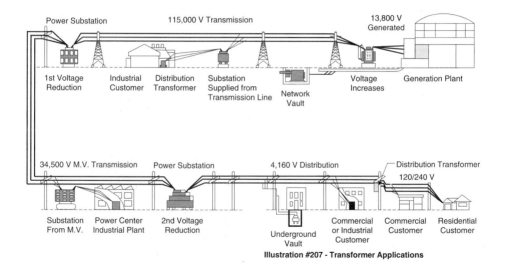

Illustration #207 - Transformer Applications

Operating Principles (cont'd)

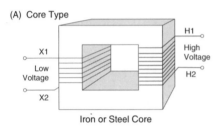

(A) Core Type

Iron or Steel Core

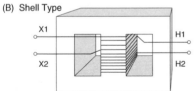

(B) Shell Type

Notes: 1. Iron or Steel Core Material has Low Retentivity to Reduce Hysteresis Losses.

2. Laminated Core reduces Eddy Currents.

Illustration #208 - Basic Transformer Types

There are certain fundamental relationships between the primary and the secondary sides of transformers. The relationship between the magnitude of the primary voltage (Vp) and the secondary voltage (Vs) is directly related to the number of turns in the primary winding (Np) and the secondary winding (Ns). Expressed in an equation form it appears as:

$$Vp/Vs = Np/Ns$$

(Np/Ns is known as the ***turns ratio***)

The voltage of the secondary circuit may be higher than the primary circuit if the number of secondary winding turns are greater than those of the primary, and vice versa, as shown in illustration #209. The relationship between the magnitude of the primary current (Ip) and the secondary current (Is) is inversely related to the number of turns in the primary winding (Np) and the secondary winding (Ns) and may be expressed in a mathematical form.

Operating Principles (cont'd)

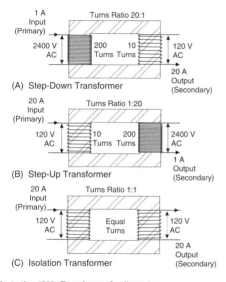

(A) Step-Down Transformer

1 A Input (Primary)

Turns Ratio 20:1

2400 V AC

200 Turns 10 Turns

120 V AC

20 A Output (Secondary)

(B) Step-Up Transformer

20 A Input (Primary)

Turns Ratio 1:20

120 V AC

10 Turns 200 Turns

2400 V AC

1 A Output (Secondary)

(C) Isolation Transformer

20 A Input (Primary)

Turns Ratio 1:1

120 V AC

Equal Turns

120 V AC

20 A Output (Secondary)

Illustration #209 - Transformer Configuration

The mathematical expression for current and the number of turns is:

$$Ip/Is = Ns/Np$$

Transformers are rated in volt-amperes or more often in kilo-volt-amperes (kVA). Loads that are connected to transformers vary in their power factor and because of this, transformers are rated in kVA instead of kW.

The kVA rating of the transformer refers to the capacity of the primary circuit as well as the capacity of the secondary circuit. As a first approximation, the kVA of the primary is equal to the kVA of the secondary and in mathematical terms is expressed as:

$$kVAp = kVAs$$

In combining all the relationships stated, it becomes apparent that if the voltage is stepped up then the current is stepped down and if the voltage is stepped down the current is stepped up. These relationships may be summarized in this mathematical form:

$$Np/Ns = Vp/Vs = Is/Ip$$

Polarity and Terminal Markings

Transformer windings are labelled or marked to indicate polarity and to designate high and low voltage sides. NEMA/EEMAC defines lead polarity as a designation of the relative *instantaneous* directions of the currents in the transformer leads (the phase relationship among the different windings). Primary and secondary leads have the same instantaneous polarity if the current enters the primary lead at the same time that the current leaves the secondary lead, as shown in illustration #210.

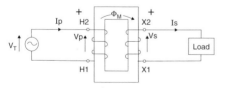

Illustration #210 - Transformer Currents

The terminal markings of power, distribution and control transformers are stamped with a letter to indicate the relative voltage level: H for high voltage and X, Y, Z in order of decreasing low voltage; and a numeral, 1, 2, 3, 4, etc. (0 indicates neutral) to indicate the relative phase relationships among the different windings.

Terminals with the same numerical markings have the same instantaneous polarity; if H1 is instantaneously positive so is X1 and if current is entering H2 then current is leaving X2, as shown in illustration #210.

The terms additive and subtractive polarity refer to standardized transformer terminal lead positions, as shown in illustration #211. The polarity of a transformer is subtractive when H1 and X1 are adjacent, and additive when H1 is diagonally located with respect to X1.

Polarity and Terminal Markings (cont'd)

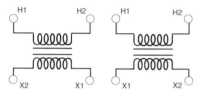

(A) Additive Polarity   (B) Subtractive Polarity

Illustration #211 - Terminal Lead Markings

Note: Polarity has no bearing on internal voltage stresses, yet subtractive polarity has a small advantage over additive polarity in voltage stresses between external leads. If two adjacent high and low voltage leads should accidentally come in contact, the voltage across the other leads would be the sum of high and low voltages for additive polarity, and their differences for subtractive polarity.

Table #72 lists the NEMA/EEMAC recommended polarity class for transformers:

NEMA/EEMAC Recommended Polarity Classes for Transformers	
Polarity Class	**Transformer**
Additive	200 kVA and smaller having high voltage windings rated 8660 V or less.
Subtractive	Over 200 kVA having high voltage windings rated 8660 V or less.
	Transformers over 8660 V regardless of their kVA size.
	Line connected instrument transformers.

Table #72 - NEMA/EEMAC Transformer Polarity Classes

Nameplate Data

Data appearing on a transformers nameplate reveals key information about the performance and limitations of the transformer.

The nameplate shown in illustration #212 is an example of what information may appear on some transformers.

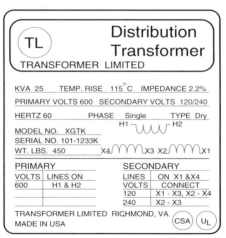

Illustration #212 - Transformer Nameplate

Nameplate Data (cont'd)

Nameplate data conforms to ANSI standards. However, at times information on the nameplate may be unique to the transformer manufacturer. Always consult with the manufacturer for complete information. The following is an explanation of the nameplate information noted in illustration #212:

1. Manufacturers name, logo and classification of transformer: Transformers may be classified as either instrument, control, distribution or power (ANSI/IEEE classifies transformers as category 1, 2, 3 or 4). Control transformers are rated in VA; instrument transformers are rated in VA burdens; distribution transformers are rated in kVA and range in sizes from 3 to 500 kVA; and power transformers are rated in kVA and MVA, and include transformers rated over 500 kVA.

2. Temperature rise: Refers to the maximum allowable difference in temperature between the transformers winding and a base ambient temperature of 40°C.

The sum of the base ambient temperature and the temperature rise equals the maximum operating value.

3. Load capacity: Rating of the transformer is expressed in kVA. The kVA of a transformer should be adequate not only for the present load but should also have enough extra capacity in the event that additional loads are added. Standard transformer kVA ratings are listed in table #73.

4. Transformer impedance: Impedance is expressed as a percentage value, and may be defined as the percentage of rated primary voltage required to cause **rated current** to circulate through the primary winding when the secondary winding is short circuited. Impedance of a transformer is one expression of the transformer's ability to limit short circuit currents. See Section Five, page 212, for more information on transformer impedance.

Nameplate Data (cont'd)

Impedance of the transformer becomes important when two or more transformers operate in parallel. If paralleled transformers do not have identical impedance values, the load that they supply will not be shared equally and this could result in one of the transformers becoming overloaded.

Percentage impedance values for transformers vary depending on transformer size and voltage rating as shown in table #74. Values shown are based on ANSI and NEMA standards.

5. Transformer voltage: Ranges from extra low voltage to high voltage. Voltages shown on the nameplate refer to the rating of the windings. Depending on the transformers design, several voltage values may appear on the nameplate. This could be due to multiple windings or phases.

Standard Base kVA Transformer Ratings			
Single-phase			
3	75	1,250	10,000
5	100	1,667	12,500
10	167	2,500	16,667
15	250	3,333	20,000
25	333	5,000	25,000
37½	500	6,667	33,333
50	833	8,333	
Three-phase			
15	300	3,750	25,000
30	500	5,000	30,000
45	750	7,500	37,500
75	1,000	10,000	50,000
112½	1,500	12,000	60,000
150	2,000	15,000	75,000
225	2,500	20,000	100,000

Table #73 - Standard Base kVA Ratings of Transformers

Nameplate Data (cont'd)

Standard Impedance Values for Three Phase Transformers		
High Voltage Rating (Volts)	**kVA rating**	**Percent Impedance**
Secondary Unit substation transformers		
2,400 - 31,800	112.5 - 225	Not less than 2.0
2,400 - 13,800	300 - 500	Not less than 4.5
2,400 - 13,800	750 - 2,500	5.75
22,900	All	5.75
34,400	All	6.25
Liquid-immersed transformers		
2,400 - 22,900		5.5
26,400 and 34,400	500	6.0
43,800	to	6.5
67,000	30,000	7.0
115,000		7.5
138,000		8.0

Table #74 - Transformer Percentage Impedances

Nameplate Data (cont'd)

6. Frequency: In North America frequency is standardized at 60 Hz.

7. Phase: Transformers are either single or three phase.

8. Types: There are four basic types of transformers to choose from. Oil filled, non-flammable insulating liquid filled (silicon-liquid), dry type and gas filled.

9. Model and serial numbers: These are coded in accordance to the manufacturers' system. Key information is often included in with these numbers. Manufacturers' manuals and catalogs will usually provide this information.

10. Wiring: Connection tables or wiring connection diagrams provide an aid to the electrician when doing the installation work.

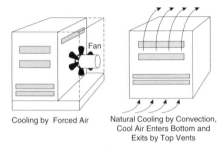

Cooling by Forced Air

Natural Cooling by Convection, Cool Air Enters Bottom and Exits by Top Vents

Illustration #213 - Ventilation of Dry Type Transformers

Insulation and Cooling

Two general types of cooling methods are used for transformers; dry or air type and liquid type. Dry type transformers may either be of the ventilated (fans are used to remove transformer heat) or the unventilated type (cooling by convection). Both are shown in illustration #213.

Insulation and Cooling (cont'd)

Fans can help dry-type transformers rated 300 kVA and above obtain up to a 33% increase in kVA ratings. These fans are controlled by relays that turn on and off in accordance with presettable thermal devices.

Liquid filled transformers have the transformer coils and core submerged in a liquid such as oil, silicone fluid, or a high molecular weight hydrocarbon. The liquid may either be naturally circulated through the transformer or it may be forced by circulating pumps. Often radiator fins are attached to the tank of a liquid filled transformer to dissipate the heat to the surrounding air. Fans may also be employed on liquid filled transformers to improve heat removal from the transformer tank.

Liquid cooled transformers from 750 kVA through 2499 kVA may obtain an additional 15% increase in their kVA rating with fan cooling, and transformers 2500 to 12000 kVA may obtain an additional 25% increase.

An effective cooling system increases the capacity of a transformer anywhere between 25% and 67%, as noted in table #76.

Table #75 lists some of the methods that are used to remove heat from transformers and the designations used to identify these methods.

Insulation and Cooling (cont'd)

Designation of Cooling Methods		
Cooling Method	**C.S.A.**	**A.N.S.I.**
Oil immersed, Natural circulation, Self cooled (Oil Natural Air Natural)	ONAN	OA
Oil immersed, Natural circulation, Water cooled (Oil Natural Water Natural)	ONWN	OW
Oil immersed, Natural circulation, Forced air cooled (Oil Natural Air Forced)	ONAF	OA/FA
Oil immersed, Forced oil, Water cooled	OFWN	FOW
Oil immersed, Forced oil, Forced air cooled	OFAF	FOA
Oil immersed, Natural circulation, Forced air cooled (second stage of forced air cooling)	ONAF/ONAF	OA/FA

Table #75 - Transformer Cooling Method Designations

Conductors that comprise the windings of these transformers are first insulated with a Class 105°C insulation material and then immersed in an insulating liquid.

The insulating liquid enhances the dielectric qualities of the windings, allowing for smaller enclosure designs at higher voltages. The fluids used in liquid filled transformers serve two key functions, cooling and insulating.

Insulation and Cooling (cont'd)

Cooling Methods and Transformer Capacity		
Type of Cooling		Loading Capacity
C.S.A.	A.N.S.I.	
ONAN	OA	100%
ONAN/ONAF	OA/FA	100/133%
ONAN/ONAF/ONAF	OA/FA/FA	100/133/167%
ONAN/ONAF/OFAF	OA/FA/FOA	100/133/167%
OFAF	FOA	167%
ONWN	OW	125%
OFWN	FOW	167%

Table #76 - Transformer Cooling Methods and Loading Capacity

Dry type transformers depend on the dielectric properties of their insulating wrap and on the medium surrounding them, which is usually air or an inert gas.

NEMA/EEMAC have designated four classes of insulation for transformer windings:

Insulation and Cooling (cont'd)

1. Class 105°C (formerly Class A): Material consists of cotton, silk, paper and other organic materials that should not be exposed to average conductor temperatures exceeding 95°C, or 55°C rise over a 40°C ambient. A hot spot allowance of an additional 10°C is allowed resulting in a maximum allowable temperature of 105°C. This type of insulation is no longer being used in dry type transformers but is commonly employed in liquid filled transformers.

2. Class 150°C (formerly Class B): Material consists of combinations of mica, glass fiber and asbestos, with bonding substances that should not be exposed to average conductor temperatures exceeding 120°C, or an 80°C rise above 40°C. A hot spot allowance of an additional 30°C results in a maximum allowable temperature of 150°C.

3. Class 185°C (formerly Class F): Material is similar to that of Class 150°C except that the bonding and impregnating is done differently. A temperature rise of 115°C over a 40°C ambient is permissible with a hot spot allowance of 30°C resulting in a maximum allowable temperature of 185°C.

4. Class 220°C (formerly class H): Material consists of silicone, elastomer, mica, glass fiber or asbestos with appropriate silicone resins. This class of material permits a rise of 150°C over a 40°C ambient with a hot spot allowance of 30°C resulting in a maximum allowable temperature of 220°C. Dry type transformers 30 kVA and over use Class H insulation. The advantage of having this insulation lies in the fact that conductors may be operated at a higher temperature and therefore may be smaller in size resulting in minimum amounts of conductor and core.

Transformer Efficiency

Transformers, being static electrical machines, operate at very high efficiencies. Efficiencies of 96% to 99% are common, with little variation from no load to full load. Losses in a transformer consist of copper losses, core losses (hysteresis and eddy currents), flux leakage losses and dielectric losses. In combination, all of these losses result in the production of heat, as shown in illustration #214, which limits the operating capacity of the transformer.

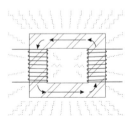

Heat = Copper Losses + Eddy Current Losses + Hysteresis Losses
+ Flux Leakage + Dielectric Absorption

Illustration #214 - Transformer Losses

The higher the efficiency of the transformer, the lower the operating costs and the lower the temperature at which it operates.

Transformer Noise

Transformer noise is caused by a phenomenon called magnetostriction, which means that if a piece of steel is magnetized it will extend itself. When the magnetization is removed, it goes back to its original condition. A transformer is magnetically excited by an alternating voltage and current so that the ferromagnetic material becomes extended and contracted twice during a full cycle of magnetization.

A transformer core consists of many layers of steel and each of these layers is being stressed magnetically at varying rates. The sum total of all these stresses produces magnetic noise. The transformer hum is caused by the extension and contraction of the core laminations as they are being magnetized. Transformer noise levels are measured in decibels (dB).

Transformer Noise (cont'd)

Because the transformer core is not symmetrical, added noise is composed of frequencies produced by the fundamental (60 Hz) and its harmonics. The difference in noise in a transformer operating at no load versus the same transformer operating at full load is usually no greater than 1 or 2 dB. The noise from a transformer is mechanical in origin and therefore vibration noise will also be present along with the audible noise. Manufacturers must meet NEMA/EEMAC noise standards for transformers and limit the audible values to those published in their standards, some of which are shown in table #77. Attenuation (reduction) of transformer noise may be accomplished by using one or more of the following methods:

1. Place the transformer in a room in which the walls, floor and ceiling will reduce the noise to a person listening on the other side and minimize sound travelling to adjacent rooms.

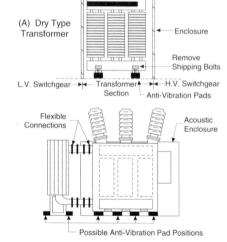

(A) Dry Type Transformer

L.V. Flexible Connection
H.V. Flexible Connection
Enclosure
Remove Shipping Bolts
L.V. Switchgear — Transformer Section — H.V. Switchgear
Anti-Vibration Pads

Flexible Connections
Acoustic Enclosure
Possible Anti-Vibration Pad Positions

(B) Liquid Filled Transformer

Illustration #215 - Transformer Core & Coil Isolation

Transformer Noise (cont'd)

Average Sound Levels (decibels)			
kVA Rating	Liquid Type (69 kV and below)	Dry Type Ventilated (15 kV and below)	Sealed
151 - 300	55	58 (67)	57
301 - 500	56	60 (67)	59
501 - 700	57 (67)	62 (67)	61
701 - 1,000	58 (67)	64 (67)	63
1,001 - 1,500	60 (67)	65 (68)	64
1,501 - 2,000	61 (67)	66 (69)	65
2,001 - 3000	63 (67)	68 (71)	66
3,001 - 4,000	64 (67)	70 (73)	68
4,001 - 5,000	65 (67)	71 (74)	69
5,001 - 6,000	66 (67)	72 (75)	70
6,001 - 7,500	67 (68)	73 (76)	71
10,000	68 (70)	--	--
12,500	69 (70)		
15,000	70 (71)		
20,000	71 (72)		
25,000	72 (73)	Note: The values in paren-	
30,000	73 (74)	theses represent the	
40,000	74 (75)	sound level when the first	
50,000	75 (76)	stage of forced cooling is	
60,000	76 (77)	in operation	
80,000	77 (78)		
100,000	78 (79)		

Table #77 - Transformer Sound Level Standards

Transformer Noise (cont'd)

2. Place the transformer inside an enclosure which uses limp wall technique. This is a method which uses two thin plates separated by a viscous (rubbery) material. The noise strikes the inner sheet and its energy is then absorbed by the viscous material. The outer sheet should not vibrate.

3. Place a screen wall around the transformer, or locate the transformer behind some natural barriers such as shrubs, etc., or if possible move the transformer far from the offending area.

4. Isolate the core and coils of the transformer from the ground by using anti-vibration pads as shown in illustration #215.

5. Make sure all connections to solid reflecting surfaces are flexible. This includes incoming cables, busbars, and stand off insulators. Solid connections from a vibrating transformer to a solid structure will transmit vibration.

6. Make sure that shipping bolts are removed so that they do not interfere with the purpose of the anti-vibration pads.

7. Lower sound levels with special designs and selecting superior magnetic materials.

Basic Impulse Level (BIL)

The basic impulse level of a transformer is an important measure of the ability of a winding to withstand transient overvoltages. If voltages in excess of those that the insulation is designed for are impressed on the winding, the insulation may be punctured and result in total winding failure.

The BIL rating of a transformer indicates that it was tested using an impulse voltage that rises to its peak value in 1.2 microseconds (μs), and then decays to 50% peak voltage after a total of 50 μs has elapsed. The transformer line terminal bushing must be assigned a BIL class as shown in table #78. Table #78 is based on IEEE and NEMA/EEMAC standards.

Basic Impulse Level (BIL) (cont'd)

Basic Impulse Insulation Level (kV) - Full Wave					
System Line-to-Line Volts	Insulation Class (kV)	Liquid Insulated		Dry Type	
		Power	Distribution	Ventilated	Gas Filled and Sealed
120-600	1.2	45	30	10	30
2,400	2.5	60	45	20	45
4,160	5.0	75	60	25	60
4,800	5.0	75	60	25	60
6,900	8.7	95	75	35	75
7,200	8.7	95	75	35	75
12,470	15.0	110	95	50	95
13,200	15.0	110	95	50	95
13,800	15.0	110	95	50	95
14,400	15.0	110	95	50	95
22,900	25.0	150	150		
23,000	25.0	150	150		
26,400	34.5	200	200		
34,500	34.5	200	200		
43,800	46.0	250	250		
46,000	46.0	250	250		
67,000	69.0	350	350		
69,000	69.0	350	350		
92,000	92.0	450			
115,000	115.0	350			
		450			
		550			
138,000	138.0	450			
		550			
		650			
161,000	161.0	550			
		650			
		750			

Table #78 - Transformer and Reactor BIL Levels

Harmonics

There are two aspects that should be considered when examining harmonics and transformers:

1. Transformer nonlinear impedance characteristic: A transformer has a non-sinusoidal magnetizing current. This complex magnetizing current waveform is composed of the fundamental (60 Hz) and multiples of the fundamental known as harmonics. Magnetizing current is about 5% of the rated transformer current while the magnitude of the harmonics is approximately 40% of the magnetizing current. Therefore at rated load the harmonic current will be only 2% of the total current.

At this level the harmonic currents primarily produce electromagnetic noise which may be induced into adjacent unshielded communication or data lines.

2. Loads supplied by the transformer: Nonlinear loads generate high levels of harmonic currents. Typical nonlinear loads include desktop computers, data processors, facsimile machines, AC variable speed drives, HID lighting, electronic ballasts, core and coil ballasts, power inverters, welders etc.

The amount of harmonics produced by a given load is expressed as a "K" factor. The larger the "K" factor the more harmonics are present. UL "K" factors are shown in table #79.

"K" factors reflect transformer load losses due to harmonics generated by the load. Harmonics are multiples of the fundamental frequency and when currents at frequencies above 60 Hz flow in a transformer winding, the following problems can arise:

Harmonics (cont'd)

a. The operating temperature of the transformer increases. Transformers are rated to operate at frequencies of 60 Hz, but when higher frequencies are present due to nonlinear loads, higher eddy currents and hysteresis losses will occur.

   If the load current is rich in 180 Hz (3rd harmonic current), eddy current losses will increase to nine times as high as compared to the fundamental frequency. Copper losses due to skin effect also increase. With increased losses the transformer will reach its maximum operating temperature even though it is supplying loads that are below its rating.

b. Transformers that are supplying wye connected nonlinear loads will have triple harmonics (3rd, 9th, 15th) flowing in the neutral.

Types of Load and K-Factors	
Load Type	K
Resistance heating	K-1
Incandescent lighting	K-1
Motors	K-1
Control and distribution transformers	K-1
Welders	K-4
Induction heaters	K-4
HID lighting	K-4
Fluorescent lighting	K-4
Solid state controls	K-4
Telecommunication equipment	K-13
Branch circuits in classrooms and healthcare facilities	K-13
Mainframe computers	K-20
Variable speed drives	K-20
Branch circuits with exclusive loads for data processing equipment	K-20

Table #79 - "K" Factors

Harmonics (cont'd)

The presence of these harmonics will increase the neutral current by as much as 1.73 times the value of the line current. This excessive neutral current can produce severe arcing, cause fires and result in extensive cable failure.

c. Harmonic currents create harmonic voltages. These harmonic voltages increase the core exciting current by forcing the core to operate closer to the nonlinear region of the iron. By forcing the core into saturation, the exciting current may increase to within 10-20% of the rated current and since exciting current contains harmonic components, increasing exciting current compounds the situation and further increases the heating of the transformer.

Transformers that are UL "K" factor rated are designed to operate in the presence of harmonic currents.

Voltage Taps

Voltage taps are used to compensate for small changes in the primary supply of the transformer, or to vary the secondary voltage level as changes in load occur. FCBN (full capacity below normal) and FCAN (full capacity above normal) designations appear on tap changing transformers. These two designations indicate that the full kVA capacity of the transformer is available when using voltage taps below or above rated voltage.

The most commonly selected tap arrangement is the manually adjustable no load type, consisting of four 2.5% taps rated above or below the nominal primary voltage rating. Usually the tap positions are changed when the transformer is deenergized. However automatic tap changing under load is available in larger transformer sizes. Usually automatic tap changers are controlled by voltage regulators set to maintain a constant voltage.

NEC/CEC Requirements

Article 450 of the NEC, or rules 26-240 to 26-266 of the CEC, should be referred to for non-utility applications of transformers when determining the following:

a. primary and secondary conductor sizes
b. primary and secondary overcurrent protection
c. locations for transformers (vaults, equipment rooms etc.)
d. grounding and bonding of transformers

Instrument Transformers

Instrument transformers may be divided into two types:

1. current (CT)
2. potential (PT)

Current transformers (CT's) have their primary connected in series with the circuit being monitored and act to transform higher circuit currents (3000, 1200, 600, 200 A etc.) to lower standardized values of either 1 or 5 A.

Potential transformers act to transform higher voltages (13.8 kV, 4.16 kV etc.) to a standardized 120 V or 69 V. Instrument transformers serve to both insulate and isolate metering and control equipment from higher line circuit values and to provide standardized values for meters and relays. Current transformers may be classified either according to their construction (toroid, bar type, wound type), or according to their application (metering or relaying with different accuracy classes).

Instrument transformer secondaries are connected to loads, called the burden, as shown in illustration #216. Instrument transformers have both a VA burden rating and an accuracy class rating. When connecting instrument transformers to metering and control instruments, polarity is extremely important to assure proper operation. Polarity marks are present on both primary and secondary terminals. They are marked with letters (H1 and X1) or with colored dots.

Instrument Transformers (cont'd)

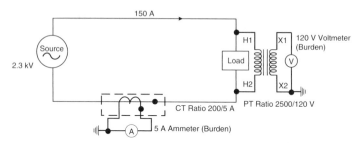

Illustration #216 - Instrument Transformer Connections

Note: The secondary winding of a current transformer must be connected to a load, or it must be short circuited at its secondary terminals (use CT shorting bars). Opening the secondary circuit while current is in the primary may cause dangerously high voltages at the secondary terminals, and may also permanently magnetize the transformer iron, thereby introducing errors into the transformer ratio.

Autotransformers

When the primary and the secondary circuits of a transformer have part of a winding in common, as shown in illustration #217, the transformer is called an autotransformer. There are distinct advantages and disadvantages to autotransformers:

Autotransformer Advantages:

a. Due to the single coil construction, auto-transformers have less leakage flux, less copper, less iron, less weight and take up less space than conventional two winding transformers.
b. Less expensive than conventional transformers.
c. Higher operating efficiency.
d. Better voltage regulation.
e. Smaller magnetizing current.

Autotransformer Disadvantages:

a. No electrical isolation between the primary and secondary circuits. Disturbances on the primary are transmitted to the secondary.
b. Lower impedance in the transformer results in higher let-through short circuit currents.

c. If the secondary winding of a step-down autotransformer should open circuit, the full primary high voltage would be impressed on the secondary load.
d. Abnormal voltages may occur in three-phase wye connected autotransformers due to either line grounds, third harmonics or line transients.

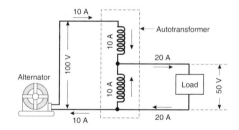

Illustration #217 - Step-Down Single Phase Autotransformer

Autotransformers (cont'd)

In a conventional transformer all of the kVA is transformed from the primary to the secondary, whereas in an autotransformer only a portion of the total kVA is transformed; the remainder is supplied directly from the primary lines to the secondary lines without transformation, as shown in illustration #218.

Note: NEC articles 210-9, 410-78, and 430-82 and CEC rule 26-264 place restrictions on the use of autotransformers when supplying interior wiring systems.

Transformer Connections

Transformer connections that are commonly encountered are either single phase or three phase. Before making any electrical connections to a transformer or group of transformers the following checks should be made:

1. Check and ensure the rating of the primary winding matches that of the source voltage.

2. Check and ensure the voltage rating of the secondary winding matches that of the load.

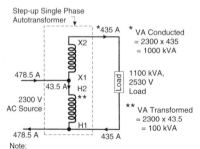

Step-up Single Phase Autotransformer

*435 A
* VA Conducted = 2300 x 435 = 1000 kVA

478.5 A

1100 kVA, 2530 V Load

43.5 A

2300 V AC Source

** VA Transformed = 2300 x 43.5 = 100 kVA

478.5 A   435 A

Note:
Transformer is Rated 100 kVA but Supplies a 1100 kVA Load

Illustration #218 - Autotransformer Operation

Transformer Connections (cont'd)

3. Check and ensure the capacity (kVA) of the transformer is greater than or equal to that of the load which it will supply.

4. Check and ensure the nature of the load connected to the transformer will be suitable for that transformer.

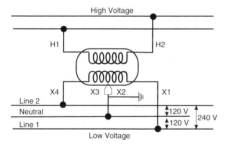

Illustration #219 - Single Phase Dual Winding Transformer

Single phase dual voltage connections

are very common for most residential single family dwellings. Transformer connections shown in illustration #219 make use of a dual winding secondary. Transformers connected as shown in illustration #219 should be connected to loads that are nearly equally balanced between the two windings.

Three-Phase Connections

Three-phase services may be either 3 wire or 4 wire. The 3 wire service is commonly called a delta service and consists of three phase conductors with no neutral. The 4 wire service is commonly called a wye service and consists of three phase conductors and one neutral.

The neutral conductor is usually grounded at the service for 208 Y/120 V and 480 Y/277 V systems. The neutral is also grounded at the secondary of the supply transformer.

Transformer Connections (cont'd)

Grounded electrical services can also be derived from delta or open delta connected transformers. These services are rated at 240/120 V, 3 phase, 4 wire with the center tap of one of the secondary windings being grounded.

There are advantages and disadvantages in using two or three single-phase transformers to obtain a system configuration as compared to using one three-phase transformer:

Advantages of the three-phase transformer:

a. lower cost
b. higher efficiency
c. less floor space and less weight
d. simplified field wiring
e. lower installation costs
f. lower shipping charges

Disadvantages of a three-phase transformer unit:

a. higher cost to stock spare units
b. greater disruption in operations when breakdown occurs
c. higher repair costs
d. difficulty in obtaining multiple taps
e. self cooled units have reduced capacities

Note: Using a unit three-phase transformer has significant advantages over using three single-phase units. Its disadvantage is primarily in the area of breakdowns, which are occuring less often as the technology improves.

For three-phase transformers or banks of three single-phase transformers, the transformer windings are usually connected either phase to phase or phase to neutral. The two most common configurations are delta and wye and the primary and secondary windings may be connected in any combination desired. ANSI combinations and connection characteristics are noted in table #80.

Transformer Connections (cont'd)

Application Characteristics of Connections									
Primary Connection	△			⅄			⅄̲		
Secondary Connection	△	⅄	⅄̲	△	⅄	⅄̲	△	⅄	⅄̲
Suitable for Ungrounded Service	Yes	Yes	Yes	Yes	Yes	Yes	(2)	No	No
Suitable for 4-wire Service	No	No	Yes	No	No	No	No	No	Yes
Phase Shift High Voltage to Low Voltage (based on standard connections and phasor diagrams)	0°	-30°	-30°	-30°	0°	0°	-30°	0°	0°
Subject to Primary Grounding Duty	No	No	No	No	No	No	Yes	No	No
Triple Harmonic Primary Line Currents Suppressed	Yes	Yes	Yes	Yes	Yes	Yes	Yes	No	No
Problem connections which should receive special consideration to avoid misapplication						(1)	(2)		(3)

See page 398 for notes (1), (2) and (3)

Table #80 - ANSI Transformer Connection Characteristics

Transformer Connections (cont'd)

(1) This connection cannot furnish a stabilized neutral. Its use may result in a phase-to-neutral overvoltage on one or two legs as a result of unbalanced phase-to-neutral load.

(2) Connections designated act as primary grounding transformers and should not be used unless intended for such duty.

(3) The 3-phase transformer of 3-legged core form construction is susceptible to tank heating with unbalanced phase-to-neutral loads or ground faults.

A standard connection method for three phase transformers has been established by ANSI. In a standard delta-delta, open-delta open-delta, and wye-wye connection, the phase displacement of H1 and X1 is 0° and in a standard wye-delta or delta-wye connection, the phase displacement of H1 leads that of X1 by 30°. Illustration #220 shows the ANSI standard connections for commonly connected three phase transformers.

Note: It is preferable to have at least one delta-connected winding in a three phase transformer bank.

Table #80 Notes

The delta connection furnishes a path for the flow of the third harmonic current in the external circuit. Delta-wye, wye-delta, and delta-delta all remove any tendency to produce triple frequency voltages. However, wye-wye connections, introduce third harmonic voltages and current asymmetry between lines and neutral that may, under certain conditions, subject the system to dangerous overvoltages (open-delta open-delta connections also give rise to third harmonic voltages).

Transformer Connections (cont'd)

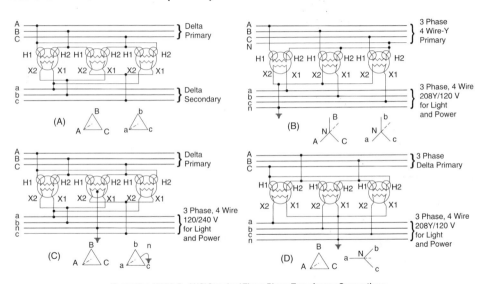

Illustration #220A-D - ANSI Standard Three-Phase Transformer Connections

Transformer Connections (cont'd)

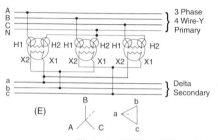

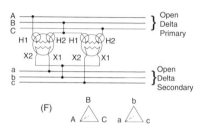

Illustration #220E,F - ANSI Standard Three-Phase Transformer Connections

Note: The open-delta open-delta connection is usually only used on distribution and power transformers when one of three single-phase delta-delta connected units fails and must be removed. This is usually an emergency expedient, and a temporary solution would be to reconnect the remaining two units in open-delta.

However, the full line current flowing in the open-delta units will be out of phase with the transformer voltages. The normal capacity is reduced to 57.7% of the three phase delta rating, and to 86.6% of the rating of the two open-delta transformers.

Transformer Connections (cont'd)

Open-delta connections also give rise to voltage unbalances. Instrument potential transformers are commonly connected in open-delta.

The following voltage and current relationships exist for the connection types noted:

1. Wye connection: $I_{LINE} = I_{PHASE}$

   $V_{LINE} = 1.73 \times V_{PHASE}$

2. Delta connection: $I_{LINE} = 1.73 \times I_{PHASE}$

   $V_{LINE} = V_{PHASE}$

3. Open-delta connection: $I_{LINE} = I_{PHASE}$

   $V_{LINE} = V_{PHASE}$

Paralleling Transformers

If greater capacity is desired, two transformers may be connected in parallel. Before two transformers can be connected in parallel the following conditions must be met for single phase (conditions 1 to 6) and three phase (conditions 1 to 7) transformers:

1. The voltage ratings of the primaries and the secondaries must be identical.
2. Proper polarities must be observed.
3. Tap settings must be identical.
4. Percentage impedance of one is between 92.5% and 107.5% of the other.
5. The equivalent resistances and equivalent reactances should have the same X/R ratio to avoid circulating currents and operation at a different power factor.
6. Frequency ratings must be identical.
7. The angular phase displacements of the two transformer banks must be the same.

Paralleling Transformers (cont'd)

Illustration #221A shows two single-phase transformers connected in parallel, and illustration #221B displays two banks of three phase transformers connected in parallel.

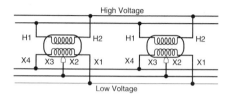

Illustration #221A - Two Single-Phase Transformers in Parallel

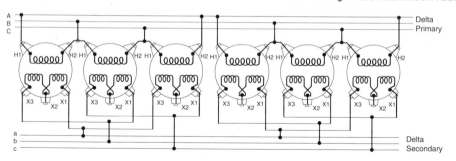

Illustration #221B - Two Banks of Three-Phase Transformers in Parallel

Paralleling Transformers (cont'd)

Table #81 provides a list of transformer connections that may be paralleled with each other.

Paralleling Configurations
0° Phase Displacement
WYE-WYE to WYE-WYE
DELTA-DELTA to DELTA-DELTA
WYE-WYE to DELTA-DELTA
DELTA-DELTA to OPEN-DELTA
OPEN-DELTA to OPEN-DELTA
WYE-WYE to OPEN-DELTA
30° Phase Displacement
WYE-DELTA to WYE-DELTA
DELTA-WYE to DELTA-WYE
WYE-DELTA to DELTA-WYE
DELTA-WYE to WYE-DELTA
Note: Transformers with 0° phase displacement cannot be paralleled with those having a 30° phase displacement

Table #81 - Allowable Parallel Transformer Configurations

Transformer Maintenance

Moisture, temperature and detrimental environments are the primary enemies that can destroy a transformer. Neglecting routine maintenance and inspections may lead to serious transformer failures and expensive power outages.

It makes good economic sense to initiate a routine maintenance program for transformers and associated auxiliary equipment. The simplest and least expensive routine maintenance procedure that can be performed on all types of transformers is to keep them clean. However, both the dry-type and liquid-filled type transformers require additional care and maintenance.

Dry Type Transformer Maintenance

Dry type transformers should be located indoors in dry areas. Care should be taken to prevent water from leaking or splashing onto the transformer. Air circulation around the transformer is important to facilitate heat removal. Therefore the transformer should be located away from walls and heat producing appliances.

Corrosive fumes break down insulation on the windings of transformers, and dust settling on transformers will reduce their heat rejection capabilities thus leading to thermal damage. Transformers should be placed in rooms or areas where they are protected against corrosive fumes and dust.

Checks should be made on the condition of the transformer's enclosure, and if rust or metal corrosion is evident, attention to restoring the original finish should be given.

When inspecting the exterior of the transformer carefully check the condition of cabling, connectors and bushings. Look for signs of discoloration, cracks or breaks in insulators and bushings.

As transformers operate with low levels of vibration, check for mechanical tightness of all external parts. Check that all gaskets, covers, and sealed openings are in good condition as too much moisture entering a transformer can cause damage. Check that all ventilation openings are clear and free of any obstructions and that the ambient temperature of the area has not increased appreciably. Periodically (depends on the degree of environmental contamination, however once a year would be recommended) either vacuum, wipe with a cloth, or use compressed air on the exterior of the transformers enclosure to remove dust and dirt. Dust and dirt can also collect and settle on the windings and core, further restricting air flow and adding to transformer heating.

Dry Type Transformer Maintenance (cont'd):

The following maintenance procedures are recommended for dry type transformers:

1. De-energize the transformer and disconnect the primary and secondary windings. **Remember to properly tag and lock out all power sources** associated with the equipment.

2. Clean the exterior of the enclosure and all connected parts.

3. Clear ventilation openings and replace filters. Some dry type transformers may have forced air cooling. Fan motors and related control devices need to be checked regularly, and should be cleaned periodically.

4. Clean the transformer core and coil using a vacuum cleaner.

5. Clean terminal boards and cable supports by brushing with a soft bristle brush, and/or wiping with a clean, dry, soft and lint-free cloth.

6. Inspect the core and coils for signs of overheating, and for tracking or carbonization on insulating surfaces.

7. Inspect and tighten busbar connections and busbar supports.

8. Perform insulation resistance tests with a megohmmeter; measure the resistance between windings, and the resistance between windings and ground, as shown in illustration #222. If insulation resistance tests indicate that the windings have absorbed moisture or if the windings have been subject to unusually damp conditions, they should be dried by external heat, internal heat, or by a combination of the two.

9. Perform a turns ratio test with a TTR test set.

Dry Type Transformer Maintenance (cont'd)

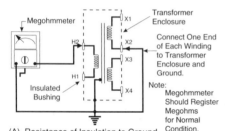

(A) Resistance of Insulation to Ground

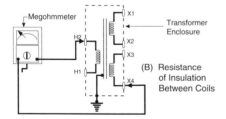

(B) Resistance of Insulation Between Coils

Illustration #222 - Resistance Tests With a Megohmmeter

Liquid Filled Transformer Maintenance

Liquid filled transformers require more maintenance and attention than dry type transformers. Since many if not most liquid filled transformers are located outdoors, they are exposed to more environmental contaminants. The surface of porcelain bushings should be periodically cleaned (once a year is recommended). This can be done manually by using a soft cloth, soap and water. Where environmental conditions are less severe these cleanings may be unnecessary, but regular inspections of the bushings should be made and any cracks, tracking, or discoloration noted and investigated.

Examine the fluid level of the transformer and look for any possible leaks. Most liquids used in transformers have a low evaporation rate; and if the liquid level is low the most probable reason would be a leak due to tank rupture, or the transformer becoming so hot that oil has boiled out.

Liquid Filled Transformer Maintenance (cont'd)

Discoloration of the liquid and the presence of sludge may indicate a high degee of oxidation. The accumulation of sludge on transformer coils and in the cooling ducts reduces heat transfer capability, resulting in higher operating temperatures. If the transformer is badly sludged, the liquid should be drained and the inside of the transformer hosed down with a clean liquid. The old liquid should then be returned to the transformer tank through a filter press.

Moisture in the liquid may be detected by testing samples of the fluid at high voltage to determine its dielectric strength. Transformer insulating oils with more than 80 ppm of dissolved water should be cause for concern and corrective action should be taken. If moisture is detected, all gaskets and bushings should be checked for points of possible entry.

Temperature checks should be made regularly. If unusually high temperatures are evident, an investigation and corrective action must be taken. A periodic test (once a year is recommended) for the presence of combustible gases in the liquid should be done. Some liquids when heated give off gases which will ignite and explode. The cause of the heating which produces these gases can be determined by sampling the gas and obtaining an analysis of its composition, as shown in table #82 (combustible gases in concentrations over 1% are a cause for concern and merit further investigation).

A summary of maintenance tasks that may be performed on liquid filled transformers are listed in table #83A and #83B. A well planned maintenance program begins with good documentation regarding each and every transformer. Every inspection and every scheduled maintenance should be recorded and subsequent findings compared to previous ones.

Liquid Filled Transformer Maintenance (cont'd)

Combustible Gas Analysis of Transformers	
Detected Gases	**Interpretations**
Nitrogen with 5% or less oxygen content.	Normal.
Nitrogen with more than 5% oxygen content.	Check tightness of transformer tank.
Nitrogen, carbon dioxide and carbon monoxide.	Transformer overloaded or operating hot, causing some cellulose breakdown. Check operating conditions and environment.
Nitrogen and hydrogen.	Corona discharge, electrolysis of water or rust.
Nitrogen, hydrogen, carbon dioxide and carbon monoxide.	Corona discharge involving cellulose or severe overloading of transformer.
Nitrogen, hydrogen, methane with small amounts of ethane and ethylene.	Arcing or low level corona discharge causing some breakdown of oil.
Nitrogen, hydrogen, methane with carbon dioxide, carbon monoxide and small amounts of other hydrocarbons. No acetylene.	Arcing or low level corona discharge in the presence of cellulose.
Nitrogen with high hydrogen, methane, ethylene and acetylene.	High energy arc causing rapid deterioration of oil. Brazed connections or turn-to-turn short circuits.
Same as the one above except carbon dioxide and carbon monoxide are present.	Same as the one above except arcing in combination with cellulose.

Table #82 - Combustible Gas Analysis of Liquid Filled Transformers

Liquid Filled Transformer Maintenance (cont'd):

Maintenance Schedule for Liquid Filled Transformers	
Every month	**Every 6 months**
Check annunciator for flags, initiate action based on findings, and walk through yard listening for unusual transformer noises.	Check equipment for maintaining 0.5 psig pressure in the tank and record N_2 cylinder pressure.
	Record and reset the maximum top oil temperature gauges.
Record and reset the maximum raise and lower taps on the load tap changer.	Record and reset the indicated and maximum hot spot winding temperatures.
	Record liquid oil levels in main unit and tap changer compartments.
Clear load tap changer breather.	Record the number of tap changer operations.
Check HV bushing liquid levels.	Test oil dielectric, color, acidity and interfacial tension.

Table #83A - Liquid Filled Transformer Maintenance Schedule

Maintenance Schedule for Liquid Filled Transformers

Each Year

Perform gas-in-oil analysis and record values (repeat more often if required) - see Table #82 for analysis.

Operate and check cooling equipment (lubricate fan motors (if required).

Megohmmeter all windings and record values.

Check tank ground connections.

Inspect for oil or gas leaks (plan corrective action if required).

Clean bushing and lightning arrester porcelains.

Make internal inspection on tank top. Look for signs of damage or loose parts (purge gas blanket with N_2 and check for less than 3% O_2).

Tighten electrical connections (torque values as per manufacturer's instructions).

Touch up paint where needed.

Manually and electrically operate load tap changer through all positions and check for binding/sticking.

Inspect voltage regulating relay and check settings.

Every 3 years

Power factor test bushings and windings, record values and report unfavorable trends.

Perform turns-ratio-test (TTR) on all windings with all no-load taps, record values.

Test sudden pressure relay and compare the results with published curves.

Check pressure relief micro-switch operation and relaying.

Bench test pressure relief for proper nameplate operation.

Every 3 years or 75,000 operations

Internal inspection of load tap changer.

Drain oil, flush with oil and filter oil as required.

Inspect: Motor drive seal.

   Raise and lower limit switches, nuts, bolts, washers and pinions.

   Mechanical problems.

   Insulation degradation.

   Electrical connections.

   Transfer switch contacts.

   Reversing switch contacts.

   Selector switch contacts.

Table #83B - Liquid Filled Transformer Maintenance Schedule

SECTION
NINE
GROUNDING

Grounding Purpose

The importance of proper grounding of electrical systems and electrical distribution equipment cannot be overstated. Under normal conditions, an electrical system can continue to operate even without proper grounding. However it is not until an abnormal condition has occured that the importance of good and proper grounding becomes evident.

There are five principle reasons for **grounding systems and circuits:**

1. To limit the voltage rise on the system due to a lightning strike.
2. To limit the voltage rise due to line switching surges.
3. To limit the voltage rise as a result of accidental contact with a higher system voltage.
4. To stabilize the voltage to ground (prevent floating neutral condition) during normal operations.
5. To facilitate the operation of overcurrent devices (circuit breakers, fuses, protective relays) in the case of a ground fault on a solidly grounded system, or to alarm and/or trip a protective device on impedance grounded systems.

There are three principle reasons for **grounding or bonding conductive** materials enclosing electrical conductors or **equipment:**

1. To limit the potential difference between the equipment enclosure and ground (earth).
2. To facilitate the operation of overcurrent devices (circuit breakers, fuses, protective relays).
3. To drain leakage and static currents to ground (reduces electrical noise interference for sensitive electronic equipment and reduces spark hazard in explosive atmospheres).

Grounding Purpose (cont'd)

The five reasons for grounding systems and circuits and the three reasons for bonding equipment may be summarized by these two *grounding/bonding purposes:*

1. To protect maintenance personnel and the public from all exposed surfaces of electrical apparatus and to insure that the portions of equipment that people can contact will not be at a voltage value higher than ground potential.
2. To protect electrical equipment from disturbances that may interfere with its normal operation or result in equipment failure.

Terminology

The following terms are commonly used when describing, discussing and examining grounding and therefore need to be understood and defined (these definitions are based on NEC, CEC, IEEE and ANSI standards):

Bonding: A low impedance path obtained by permanently joining all non-current carrying metal parts to assure electrical continuity, and having the capacity to conduct safely any current likely to be imposed on it. This definition emphasizes the permanence of the connection, the positive continuity of the connection, and the ampacity to conduct fault current. This term most often applies to electrical equipment being connected so that no potential exists between the equipment and earth, as shown in illustration #223.

Bonding conductor (jumper): A reliable conductor which connects the non-current carrying parts of electrical equipment, raceways, or enclosure to the service equipment, or to the system grounding conductor or between two portions of electrical equipment, as shown in illustration #223.

Terminology (cont'd)

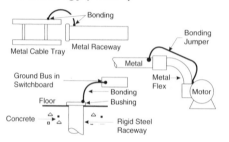

Illustration #223 - Bonding of Electrical Equipment

Counterpoise: An artificial (man-made) ground electrode consisting of a low resistance conductor which is a minimum of 200 feet (60 m) in length and laid out in a radial fashion and placed in the ground and covered. This type of ground electrode is confined to locations having high resistance soils, such as sand or rock, or where other grounding methods are found to be unsatisfactory.

Equipment grounding conductor: See definition for bonding conductor.

Ground: A conducting connection, whether intentional or accidental, between an electrical circuit or equipment and the earth, or to some conducting body that serves in place of the earth and being at the potential of earth.

Grounded: To be effectively connected with the general mass of the earth through a grounding path of sufficiently low impedance and having an ampacity sufficient at all times to prevent any current in the grounding conductor from causing a harmful voltage to exist between it and earth.

Grounded conductor: A system or circuit conductor that is intentionally grounded (eg. neutral conductor of a three-phase four-wire system is intentionally grounded) as shown in illustration #224.

Terminology (cont'd)

Grounded system: A system of conductors in which at least one (usually the middle wire or neutral point of a transformer or generator winding) is intentionally grounded, either solidly (illustration #224) or through an impedance, as in illustrations #226 and #227.

Grounding (electrode) conductor: A conductor that is used to connect service equipment or the grounded circuit of a wiring system to a grounding electrode or electrodes, as shown in illustration #224.

Grounding electrode: A means of making intimate contact with the ground (earth) and may be accomplished by natural means (buried metal water piping system, metal building frameworks, well casings, steel piling, and any other underground metal structure installed for purposes other than grounding), or by installing man-made artificial electrodes such as ground rods, buried strips of cable, grids, buried plates and counterpoises, as shown in illustration #225.

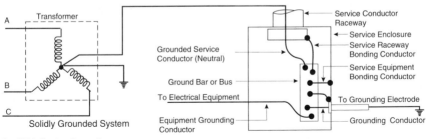

Illustration #224 - System Grounding

Terminology (cont'd)

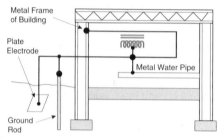

Illustration #225 - Grounding Electrodes

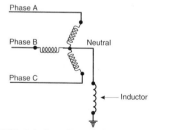

Illustration #226 - Inductance Grounded

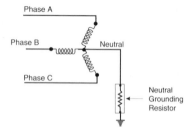

Illustration #227 - Resistance Grounded

Inductance grounded: Grounded through an inductive element, as shown in illustration #226. The reduction in ground fault current is limited to about 25% of the three phase short circuit current.

Resistance grounded: Grounded through a resistance that is connected between the electric system neutral and ground. See illustration #227.

Terminology (cont'd)

Ungrounded system: A system without an intentional hard wire connection between any of the current carrying conductors and ground. See illustration #228.

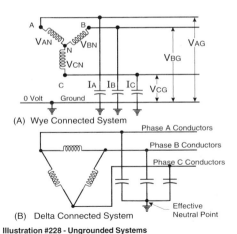

(A) Wye Connected System

(B) Delta Connected System

Illustration #228 - Ungrounded Systems

System Grounding

There are four classes of grounded electrical systems:

a. ungrounded
b. resistance grounded
c. inductance grounded
d. solidly grounded

Before choosing which grounding class to use it is necessary to recognize the limitations and advantages of each one and then compare that to local code and plant operation requirements.

Ungrounded AC system: The principal reason for the use of an ungrounded system (shown in illustration #228) is that a single ground fault on the system does not require that any part of the system be shut down. Since no hard connection has been made between the earth and any part of the system, if one of the conductors should fault to ground, there is no return path for the current.

System Grounding (cont'd)

Therefore continuity of operations is maintained in spite of one conductor being faulted to ground. While this advantage seems attractive there are some disadvantages. For the duration of the fault on the one conductor, the other two phase conductors are subjected to a *73% rise in voltage* as referenced to earth. This higher voltage stresses not only the conductor's insulation but also the insulation of the loads connected to the system.

Because of the capacitive coupling to ground, the ungrounded system is also vulnerable to *dangerous overvoltages* (five times normal or even more) caused by intermittent contact ground faults or by high inductive reactances being connected from one line to ground and attaining resonance. Special and costly equipment is needed to monitor ungrounded systems and to locate the ground faults.

Note: Conductor insulation thickness for a particular voltage is determined by the length of time that a phase to ground fault is allowed to persist. Three thicknesses have been specified:

a. 100% level....clearing time for the fault will not exceed 1 minute.

b. 133% level...clearing time for the fault will exceed 1 minute but not exceed one hour.

c. 173% level....clearing time for the fault will exceed one hour.

The 100% level may be used on any system whether solidly or low resistance grounded as long as the specified time has been met. The 133% and 173% levels apply mainly to ungrounded and high resistance grounded systems.

Resistance grounded AC system (see illustration #227): A resistor is connected between the neutral and ground to limit ground fault current to a value equal to or greater than the capacitive charging current of the system.

System Grounding (cont'd)

Typical ground fault currents range between 2-10 A for **high resistance** grounding and 400-2000 A for **low resistance** grounding.

The main purpose of **high resistance** grounding is to avoid automatic tripping of the faulted circuit at the first ground fault occurrence. This method of grounding is suitable for three-wire three-phase systems and for two-wire single-phase systems. The grounding equipment consists of a neutral resistor for wye systems and a neutral resistor and grounding transformer for a delta system. This type of grounding system limits the severe transitory overvoltages associated with ungrounded systems. Transient overvoltages are now limited to 250% of normal voltage as compared to 500% or greater in ungrounded systems.

One disadvantage to high resistance grounding is a 73% rise in voltage between the two unfaulted conductors of the three phase system and ground.

It is recommended that high resistance grounding should be restricted to 5 kV class systems with charging currents of about 5.5 A or less and should not be attempted on 15 kV systems.

The following system voltages may be **low resistance** grounded:
480 V, three-phase three-wire
600 V, three-phase three-wire
2400 V, three-phase three-wire
4160 V, three-phase three-wire
6900 V, three-phase three-wire
13,800 V, three-phase three-wire

Inductance grounded (see illustration #226): Inductance grounded systems are rarely used in industrial power systems and are not considered as an alternative to resistance grounding in medium voltage systems.

Inductance grounding is usually restricted to low voltage (under 600 V) generators. Overvoltages are minimized by selecting a reactor that limits the ground fault current to 25% of the three-phase fault current.

System Grounding (cont'd)

Solidly grounded system (see illustration #229): Solidly grounded systems exercise the greatest control of overvoltages and have the fastest response to isolating the faulted circuit. Ground fault relaying may also be used to protect against low level arcing faults and thereby protect against destructive low level arc currents. The principal disadvantage to a solidly grounded system is the high ground fault currents that occur during a ground fault.

The typical areas of application are industrial and commercial services supplying loads that affect public welfare and safety. Solidly grounded systems are used extensively at operating voltages of 600 V and less and at voltages above 15 kV.

Solidly grounded systems are most commonly used for interior wiring operating at 600 V, AC or less.

The majority of grounded systems use a wye configuration with the neutral grounded, as shown in illustration #229 (see illustration #230 for four wire delta with a high leg). The grounded conductor is usually the neutral and will be considered as such in this section.

NEC article 250 and CEC section 10 mandate the requirements for system grounding. Systems used to supply phase to neutral loads for the following voltages and configurations must be solidly grounded:

1. 120/240 V, single-phase three-wire
2. 208Y/120 V, three-phase four-wire
3. 240delta/120 V, three-phase four-wire
4. 480Y/277 V, three-phase four-wire
5. 600Y/347 V, three-phase four-wire

Note: Some exceptions do apply. See NEC/CEC.

System Grounding (cont'd)

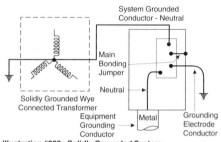

Illustration #229 - Solidly Grounded System

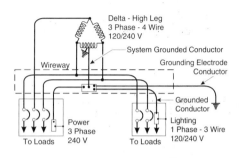

Illustration #230 - Four Wire Delta Grounded

Grounding Electrical Conductors:

The Code requires that the grounded conductor be connected to the grounding electrode system by means of the grounding electrode conductor, as shown in illustration #229. The grounding electrode conductor must be connected to the grounded conductor whenever a grounded system is used.

Note: NEC Section 384-3(e)(f) states the high leg shall be phase "B" and identified with an orange color.
CEC Rule 4-036(4) states the high leg shall be phase "A" and identified with a red color.

System Grounding (cont'd)

The following conditions and requirements apply to the grounding electrode conductor:

1. The conductor must be made of either copper, aluminum, or copper-clad aluminum (in Canada grounding electrode conductors must be copper).

Note: The following restrictions apply to aluminum or copper-clad aluminum grounding electrode conductors:

    a. May not be used where in direct contact with masonry.

    b. May not be used where in direct contact with earth.

    c. May not be used where subjected to corrosive conditions.

    d. May not be within 18 inches (450 mm) of earth when installed outdoors.

2. The conductor material must resist corrosive conditions or be protected from corrosion by being insulated.

3. The conductor may either be solid or stranded.

4. The conductor may be insulated, bare or covered.

5. The conductor must be protected from mechanical damage.

Note: When metal enclosures (conduit etc.) are used for the mechanical protection of the grounding electrode conductor, the following precautions must be taken:

    a. The metal enclosure (usually a raceway) must be electrically continuous from cabinets or equipment to the grounding electrode. The reason for this measure lies in the principle that ground fault current will produce magnetic lines of force which in turn will induce a voltage into the metal enclosure causing the metal to become very warm or hot.

    b. The metal enclosure or raceway must be fastened securely to the ground clamp or fitting.

System Grounding (cont'd)

   c. A metal enclosure that protects only a portion of the conductor's length must be bonded at each end and the bonding jumper must be the same size as the grounding conductor.

6. The conductor must be continuous throughout its length unless approved means (as shown in illustration #231) are used for splicing.

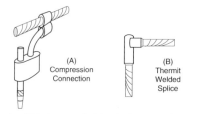

Illustration #231 - Approved Splicing for Grounding Conductor

(A) Compression Connection

(B) Thermit Welded Splice

The size of the grounding electrode conductor for either a grounded or ungrounded system shall be as per table 250-94 of the NEC (USA only) and reproduced here in table #84, and as per Table 17 for grounded systems or Table 18 for ungrounded systems of the CEC (Canada only) and reproduced here in tables #85 and #86.

Notes for Table #84

1. For both grounded and ungrounded systems if the electrode grounding conductor is connected to a made electrode (ground rod) the largest conductor required is a #6 copper or #4 aluminum.

2. For both grounded and ungrounded systems if the electrode grounding conductor is connected to a concrete-encased electrode the largest conductor required is a #4 copper.

3. For both grounded and ungrounded systems if the electrode grounding conductor is connected to a ground ring (counterpoise) the conductor is not required to be larger than the ground ring conductor (minimum #2).

System Grounding (cont'd)

Grounding Electrode Conductor (USA)			
Largest Service-Entrance Conductor or Equivalent Area for Parallel Conductors		**Size of Grounding Electrode Conductor**	
Copper	**Aluminum or Copper-Clad Aluminum**	**Copper**	**Aluminum or Copper-Clad Aluminum**
2 or smaller	1/0 or smaller	8	6
1 or 1/0	2/0 or 3/0	6	4
2/0 or 3/0	4/0 or 250 kcmil	4	2
Over 3/0 thru 350 kcmil	Over 250 kcmil thru 500 kcmil	2	1/0
Over 350 kcmil thru 600 kcmil	Over 500 kcmil thru 900 kcmil	1/0	3/0
Over 600 kcmil thru 1100 kcmil	Over 900 kcmil thru 1750 kcmil	2/0	4/0
Over 1100 kcmil	Over 1750 kcmil	3/0	250 kcmil

Note: See page 423 for table #84 notes

Table #84 - Electrode Grounding Conductor (USA only)

System Grounding (cont'd)

Grounding Electrode Conductor for Ungrounded Systems (Canada)			
Ampacity of Largest Conductor or Equivalent Multiple Conductors	Size of Grounding Conductor		
	Copper Wire	Metallic Conduit	Electrical Metal Tubing
Amps	AWG	Inches	Inches
60	8	0.75	1.00
100	8	1.00	1.25
200	6	1.25	1.50
400	3	2.50	2.50
600	1	3.00	4.00
800	0	4.00	4.00
Over 800	00	6.00	-

Table #85 - Grounding Electrode Conductor for Ungrounded Systems (Canada only)

System Grounding (cont'd)

Grounding Electrode Conductor for Grounded Systems (Canada)	
Ampacity of Largest Conductor or Equivalent Multiple Conductors	Size of Copper Grounding Conductor AWG
100 or less	8
101 to 125	6
126 to 165	4
166 to 200	3
201 to 260	2
261 to 355	0
356 to 475	00
Over 475	000

Table #86 - Grounding Electrode Conductor For
Grounded Systems (Canada Only)

Equipment Grounding/Bonding Conductor

The equipment grounding conductor is used for grounding metal enclosures that contain current carrying conductors and metal enclosures that house electrical equipment. The principal concern and role played by the equipment grounding conductor is in the safety of personnel.

This conductor establishes a continuous path for the electric current to flow under fault conditions, while at the same time maintaining the electrical equipment at a zero potential as referenced to ground. The equipment grounding conductor also serves to reduce static electrical charges.

Equipment Grounding/Bonding Conductor (cont'd)

Equipment grounding conductors may consist of one of the following:

a. a conductor of copper, aluminum, or other corrosion-resistant conducting material

b. a busbar or steel pipe

c. a metal raceway (rigid steel, rigid aluminum, cable tray, EMT)

d. a metal-covered cable assembly (mineral insulated cable, etc.)

e. a metal electrical equipment enclosure

The size of the equipment grounding conductor shall be as per Table 250-95 of NEC (USA only) and reproduced here as table #87, or as per Table 16 of CEC (Canada only) and reproduced here as table #88.

Identification

System and equipment grounding conductors must be identified by one of the following means:

1. Conductor is covered with a continuous green-colored insulation.
2. Conductor is covered with a continuous green-colored insulation with one or more yellow stripes.
3. Conductor is bare.
4. Conductors larger than #6 AWG (USA) or #2 AWG (Canada) may be identified with a green band around their insulation. This identification must be made at every point of access to the conductor.

Equipment Grounding /Bonding Conductor (cont'd)

Equipment Grounding Conductor Sizes (USA)		
Max. Rating/ Setting of Over-current Device in Circuit ahead of Equipment, Conduit, etc.	Copper	Aluminum or Copper-clad Aluminum
Amperes	Size	Size
15	14	12
20	12	10
30	10	8
40	10	8
60	10	8
100	8	6
200	6	4
300	4	2
400	3	1
500	2	1/0
600	1	2/0
800	1/0	3/0
1000	2/0	4/0
1200	3/0	250 kcmil
1600	4/0	350 kcmil
2000	250 kcmil	400 kcmil
2500	350 kcmil	600 kcmil
3000	400 kcmil	600 kcmil
4000	500 kcmil	800 kcmil
5000	700 kcmil	1200 kcmil
6000	800 kcmil	1200 kcmil

Notes for table #87

1. When conductors are run in parallel and located in more than one raceway, the equipment grounding conductors are also to be run in parallel. Each of the paralleled equipment grounding conductors is sized according to the ampere rating of the overcurrent device ahead of the conductors that it is protecting.

2. When circuit conductors are adjusted in size to compensate for voltage drop, the equipment grounding conductor must be correspondingly adjusted in size.

3. When more than one circuit is installed in a single raceway, only one equipment grounding conductor is required for all the circuits. However the grounding conductor's size must be based on the rating of the largest overcurrent device used to protect the conductors in that raceway.

4. Flexible cords protected by overcurrent devices rated 20 A or greater must have a minimum #18 copper grounding conductor.

5. The equipment grounding conductors do not have to be larger than the circuit conductors.

6. Where the overcurrent protection is an instantaneous circuit breaker or a motor protector, the equipment grounding conductor's size is based on the size of the motor overload protection that is used.

Table #87 - Equipment Grounding Conductor (USA only)

Equipment Grounding /Bonding Conductor (cont'd)

Equipment Grounding Conductor Sizes (Canada)		
Max. Rating/ Setting of Over-current Device in Circuit ahead of Equipment, Conduit, etc.	Size of Bonding Conductor AWG	
Amperes	Copper	Aluminum
20	14	12
30	12	10
40	10	8
60	10	8
100	8	6
200	6	4
300	4	2
400	3	1
500	2	0
600	1	00
800	0	000
1000	00	0000
1200	000	250 kcmil
1600	0000	350 kcmil
2000	250 kcmil	400 kcmil
2500	350 kcmil	500 kcmil
3000	400 kcmil	600 kcmil
4000	500 kcmil	800 kcmil
5000	700 kcmil	1000 kcmil
6000	800 kcmil	1250 kcmil

Table #88 - Equipment Grounding Conductor (Canada only)

Notes for table #88

1. The size of the equipment bonding conductor in no case needs to be larger than the largest ungrounded conductor in the circuit.

2. Where circuit conductors are paralleled and located in separate raceways and an equipment bonding conductor is required, the size of the bonding conductor will be determined by dividing the size of the overcurrent device by the number of paralleled raceways and then choosing a conductor using values obtained from this table.

Connection to Earth

The earth is considered to be zero potential and connection of grounded system conductors and electrical equipment to earth should ideally be with no resistance. However as shown in illustration #232, connections to earth will have some resistance. This resistance could be due to resistance between electrode and grounding conductors, resistance between electrode and adjacent earth, and resistance of the surrounding earth.

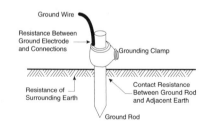

Illustration #232 - Grounding to Earth

Resistance of the electrode and resistance between the electrode and earth is very small and amounts to only fractions of an ohm. However resistance of the earth (soil) can vary between a fraction of an ohm to thousands of ohms. Remembering that the resistance value of any material is inversely proportional to its area will help in understanding illustration #233. Illustration #233 shows a series of concentric spherical shells, all of equal thickness, that radiate outward from a rod electrode. Each of these concentric lines represents a new value of resistance. The closest ring to the rod has the highest value of resistance and the farther out from the rod, the lower the resistance. The reason the resistance value is highest at the rod is because the nearest shell (concentric line) has the smallest circumferential area or cross section, and therefore has the highest resistance.

Connection to Earth (cont'd)

Successive shells outside the first have pro-gressively larger areas, and therefore pro-gressively lower resistances. There is a point where further increases in distances from the rod electrode no longer result in progressively lower resistances as the value of resistance has now effectively become zero. The first few inches away from the electrode are the most important when con-cerned with reducing the electrode resis-tance.

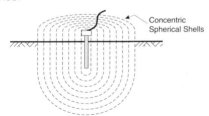

Note: 90% of the total resistance surrounding an electrode is generally within a radius of 6 - 10 feet (1.8 - 3 m)

Illustration #233 - Ground Rod Resistance

The soil resistivity (a measure of the soils opposition to the flow of current measured in ohm-metres) is a principal concern when considering the resistance between the elec-trode and earth. The resistivity of soils varies with the depth from the surface (the deeper the electrode is placed, generally the lower the resistivity), the type and concentration of soluble chemicals in the soil (different soil types have different resistivities - see table #89), the moisture content (the higher the moisture level, the lower the resistivity- see table #90), and the soil temperature (the higher the soil temperature, the lower the resistivity - see table #91).

Soil resistivity may be reduced anywhere from 15 to 90%, depending upon the kind and texture of the soil, by chemical treatment. There are a number of suitable chemicals for this purpose and they include sodium chlor-ide (salt), magnesium sulfate, copper sulfate and calcium chloride.

Connection to Earth (cont'd)

Resistivity of Soils			
	Resistivity $\Omega \cdot m$		
Soil	Average	Min.	Max.
Fills, ashes, cinders, brine waste, salt marsh	23.7	5.9	70
Clay, shale, gumbo, loam	40.6	3.4	163
Clay etc. with added sand and gravel	158	10.2	1350
Gravel, sand, stones, with little clay or loam	940	590	4580

Table #89 - Resistivity of Soils

Moisture and Soil Resistivity		
Moisture Content	Resistivity $\Omega \cdot m$	
% by weight	Top Soil	Sandy Loam
0	10×10^6	10×10^6
2.5	2500	1500
5	1650	430
10	530	185
15	190	105
20	120	63
30	64	42

Table #90 - Moisture and Soil Resistivity

Temperature and Soil Resistivity		
Temperature		Resistivity
°C	°F	$\Omega \cdot m$
20	68	72
10	50	99
0 (Water)	32	138
0 (Ice)	32	300
-5	23	790
-15	14	3300
Note: Based on Sandy Loam with 15.2% moisture		

Table #91 - Temperature and Soil Resistivity

Connection to Earth (cont'd)

Soil resistivity chemicals may be applied by placing them in a circular trench around the electrode in such a manner as to prevent direct contact with the electrode. In time as these chemicals are leached by moisture (saturating the soil around the electrode with water will accelerate this process) into the surrounding area, the effects become apparent. Periodic testing of ground resistivity should be done as chemical soil treatment is not a permanent solution and may require further applications.

Ground Rods

One of the most common *made electrodes* is the driven ground rod. These ground rods are manufactured in diameters of 3/8, 1/2, 5/8, 3/4, 1 inch (10, 13, 16, 19, 25 mm), and in lengths of 5 to 40 feet (1500 to 12000 mm) with the most common being 10 feet (3000 mm). Ground rods are made of either copperweld, copper clad steel or galvanized steel.

Note: While the effect of the ground rod's diameter on the resistance of the connection to earth is small, its length is extremely important in both lowering resistance and providing adequate area for the flow of ground fault current. Never compromise the length of a driven ground rod by cutting it off short unless a proper grounding study has determined it safe to do so. Shortening the ground rod will increase the resistance of the grounding system and could endanger life and property.

Ground Measurement

There are many formulas that have been devised to calculate the resistance of a grounding system but the only certain way to determine ground resistance is to measure it after the grounding system has been completed.

Connection to Earth (cont'd)

Various methods available to measure ground resistance make use of two auxiliary electrodes in addition to the one under test. The following ground resistance test methods are commonly used (for a complete treatment of these methods refer to Standard Handbook for Electrical Engineers by D.G. Fink and J.M. Carroll):

1. Fall of Potential Method
2. Two Point Method
3. Three Point Method
4. Ratio Method

Static Electricity Grounding

Industrial plants that handle solvents, dusty materials, or flammable products have a potentially hazardous operating condition because static charges may accumulate on equipment, materials being handled or on operating personnel.

The discharging of these electrical charges in the presence of flammables may result in an explosion and/or fire. The probability of producing an electrical static charge depends on the materials characteristic, the speed of separation, the area in contact, the motion between substances (friction) and atmospheric conditions (degree of humidity). Electrostatic charges may reach values of 100,000 V and may have a sparking distance of 10 inches (250 mm).

Static Electricity Grounding (cont'd)

A static electricity spark of this magnitude could release enough energy to easily ignite a flammable mixture. Most static problems may be solved by bonding various parts of equipment together and then bonding to the grounding system. Another often effective nonelectrical solution to the problem of static build-up is to artificially increase the humidity in the area, space or location that is prone to static.

Lightning Protection Grounding

Lightning is an electrostatic discharge between clouds or between clouds and earth. The average lightning discharge is estimated at 200 million volts, 30,000 A and travels from the cloud to earth in less than one-thousandth of a second.

The fundamental theory of lightning protection is to provide a means by which a discharge may enter or leave the earth without passing through a high impedance. Therefore the basic design principle incorporated in protecting structures and personnel from the harmful effects of a lightning strike is to provide a low impedance metallic path to ground for the lightning, or to use lightning arrestors that are connected to earth.

All metal located within a few feet of lightning conductors should be bonded to the lightning conductor to avoid arcing. Reinforcing bars in concrete construction should also be bonded to the lightning grounding conductors.

Lightning Protection (cont'd)

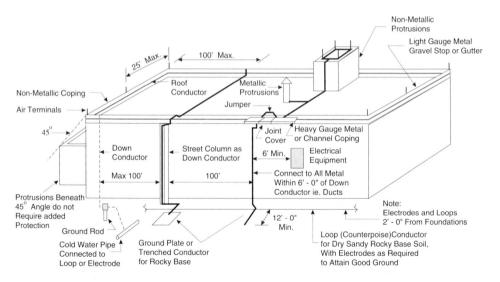

Illustration #234 - Lightning Protection System

Lightning Protection Grounding (cont'd)

A lightning protection system consists of suitable air terminals or diverter elements at the top of or around the structure to be protected, as shown in illustration #234, and connected by an adequate down conductor to the earth. The down conductor should not include any high impedance portions or connections and sharp conductor bends or loops should be avoided.

The NEC and the CEC require that electrical equipment be kept clear of lightning rod conductors (6 feet in USA and 2 m in Canada). However if that is not possible, then bonding (conductor to be minimum of #6 and same size as grounding electrode conductor) of the service grounding conductor and the lightning grounding electrode is permitted.

Agricultural Livestock Building Grounding

When cattle stand on concrete or dirt floors and come in contact with a metal surface they may feel a tingle voltage, which is actually a low voltage electrical shock. The shock may be mild enough to cause animal behavior or productivity to be altered, or severe enough to kill. One way to stop the tingle voltage is to create an equipotential plane by bonding all conductive materials in the building together, as shown in illustration #235.

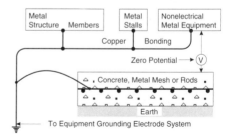

Illustration #235 - Grounding in Agricultural Building

Agricultural Livestock Building Grounding (cont'd)

Another solution to the tingle voltage problem is to use a tingle voltage filter, as shown in illustration #236. This device is connected in series with the grounding conductor and located between the neutral block and the grounding system, at the main service.

In order for the device to function, the only grounding connection allowed to the neutral on the barn panelboard is through the filter. Any other contact between the neutral and ground would bypass the filter and therefore defeat its purpose.

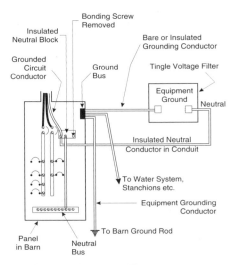

Illustration #236 - Tingle Voltage Filter

Cathodic/Anodic Protection

A brief introduction to the topic of electro-chemical corrosion control is warranted (40% of the annual steel output goes to replace products ruined by corrosion). Cathodic protection is the reduction or elimination of corrosion by making the metal a cathode by means of an impressed direct current or attachment to a sacrificial anode (usually magnesium, aluminum or zinc). As shown in illustration #237, sacrificial anodes are placed in the ground and connected to the cathodic protection system. These anodes now become the target of galvanic corrosion rather than the steel structures.

Illustration #237 also shows that direct current flows from the anodic area (called the ground bed) into the soil, through the soil and onto the cathodic areas (protected steel structure) and back through the rectifier.

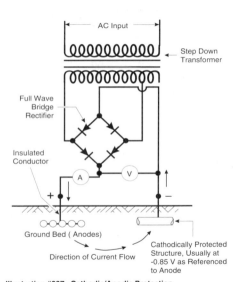

Illustration #237 - Cathodic/Anodic Protection

Cathodic/Anodic Protection (cont'd)

The galvanic potential between anode and cathode will cause a current to flow. This current is limited by such factors as the resistivity of the environment and the degree of polarization at anodic and cathodic areas. Corrosion occurs where the current discharges from the metal into the soil at anodic areas but no corrosion occurs where the current leaves the environment and enters the metal structure (cathodic area).

In applying cathodic protection to a structure, the objective is to force the entire structure surface exposed to the environment to collect current from the environment. This is accomplished by making the structure slightly more negative than its surroundings. The CEC in section 80 mandates the wiring methods required when installing Cathodic/Anodic protection systems.

SECTION TEN
LIGHTING

Lighting Terms

Visible light can be artificially created and controlled. Luminaires (lighting fixtures) change electrical energy into visible light energy and are the principal means used to produce artificial visible light. There are numerous types of artificial light sources available and each type has its own unique light producing qualities. The Illuminating Engineering Society (IES) is the principal organization in establishing and defining lighting standards and terminology. Interpreting and understanding the capabilities and qualities of different luminaires first requires an understanding of lighting terminology.

Candlepower: The intensity of a lighting source in a particular direction. The candlepower of a light source refers to the intensity or strength of the light and may be compared to voltage in an electrical circuit.

Candlepower curve: As shown in illustration #238, a candlepower curve is a graphic representation of the shape of the emitting light. The candlepower curve reveals the light pattern created by the luminaire.

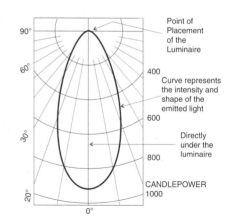

Illustration #238 - Candlepower Curve

Lighting Terms (cont'd)

Coefficient of Utilization (CU): This is a numerical representation of the effectiveness of a luminaire when installed in a particular area or space. The coefficient is a composite number representing how a luminaire will respond when installed in a room or area having a certain shape and finish. As shown in table #92, CU values range between 0 and 1, however they may also appear as percentage values, 0 - 100%. Manufacturers publish CU tables for use in calculating the average level of illumination appearing on a work plane or surface.

Color: Unique observable differences between one hue or chromaticity and ranging in variation from violet to red.

The color of light is determined by its wavelength and ranges from the shortest wavelength expressed in nanometres (1×10^{-9} m) at 380 to the longest at 780.

The color violet is at the short end of the visible spectrum, as shown in illustration #239, and red at the long end. Colors at the short end of the spectrum are perceived to be cool colors and colors at the long end of the spectrum are perceived to be warm colors. The human eye is most sensitive to the color yellow. White light contains all the colors, black light is devoid of any colors and monochromatic light contains only one color.

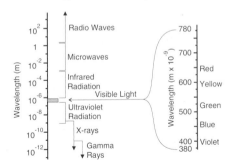

Illustration #239 - Visible Principal Colors

Lighting Terms (cont'd)

Coefficient of Utilization (SAMPLE ONLY) Based on Room Cavity Ratio (RCR) ie. the room shape								
Ceiling Reflectance	80%			70%			50%		
Wall Reflectance	50%	30%	10%	50%	30%	10%	50%	30%	10%
RCR =1	.68	.65	.63	.65	.63	.61	.61	.60	.58
2	.60	.56	.53	.58	.55	.53	.55	.52	.49
3	.54	.49	.45	.52	.48	.45	.50	.46	.43
4	.49	.43	.40	.47	.43	.39	.45	.41	.38
5	.44	.38	.34	.43	.38	.34	.40	.36	.33
6	.40	.34	.30	.39	.34	.30	.37	.32	.29
7	.36	.31	.27	.35	.30	.26	.33	.29	.26
8	.32	.27	.24	.32	.27	.23	.30	.26	.23
9	.29	.24	.21	.29	.24	.20	.27	.23	.20
10	.27	.22	.18	.26	.21	.18	.25	.21	.18

Notes:
1. CU Tables are based on a floor reflectance of 20%.
2. CU Numbers represent luminaire effectiveness in a given space or area - 1 is the optimum.
3. Spacing from centerline to centerline of luminaires not to exceed 1.2 x the mounting height.

Table #92 - Luminaire CU Values (Sample)

Lighting Terms (cont'd)

Color Rendering Index (CRI): This is a numerical representation that expresses the effect of a light source on the color appearance of objects in comparison with their color as they would appear under a reference light. Values are usually between 0 and 100, with 100 representing the highest color rendering value. The higher an artificial light source's CRI, the truer will be the colors seen under this light. Table #93 lists the CRI values of some common lamps.

Color Rendering Description	
CRI	Color Rendering
75 - 100	Excellent
60 - 75	Good
50 - 60	Fair
0 - 50	Poor

Table #93A - Color Rendering Description

Color Rendering Index Values		
Light Source	CRI	Color Rendering
Incandescent lamps	97	Excellent
Fluorescent, full spectrum 7500	94	Excellent
Fluorescent, cool white deluxe	87	Excellent
Fluorescent, warm white deluxe	73	Good
Metal halide (400 W, clear)	65	Good
Fluorescent, cool white	62	Good
Fluorescent, warm white	52	Fair
High pressure sodium (400 W, diffuse coating)	32	Poor
Low pressure sodium	-	(undefined)

Table #93B - Color Rendering Index Values

Lighting Terms (cont'd)

Color Temperature: A temperature expression in degrees Kelvin of the color of a light source. The lower the color temperature, the warmer the perceived colors and the higher the color temperature, the cooler the perceived colors. As seen in table #94, the color temperatures of artificial light sources vary from around 1700 to over 7000 degrees Kelvin. Knowing the color temperature of a light source will help in determining what kind of an atmosphere will be created.

Efficacy: This is a ratio of light output to power input (lumens/watt). This numerical expression reflects the ability of a lamp to produce light, and ranges in value from less than 10 to near 200 lumens per watt, as shown in table #95.

The higher the lumens per watt, the lower the cost in operating the lamp.

Color Rendering Index Values		
Light Source	**Color Temp.**	**Atmosphere Description**
Sky - extremely blue	25,000 K	cool
Mercury Vapor	7,000 K	cool
Sky - overcast	6,500 K	cool
Sunlight at noon	5,000 K	cool
Fluorescent - cool white	4,300 K	cool
Metal halide (400 W, clear)	4,300 K	cool
Fluorescent, warm white	3,000 K	warm
Incandescent (100 W)	2,900 K	warm
High pressure sodium (400 W, clear)	2,100 K	warm
Candle flame	1,800 K	warm
Low pressure sodium	1,740 K	warm

Table #94 - Color Temperature

Lighting Terms (cont'd)

Efficacy of Lamps (Lumens per Watt)	
Candle (Luminous Efficacy Equivalent)	0.1
Oil Lamp (Luminous Efficacy Equivalent)	0.3
Original Incandescent Lamp (1879)	1.4
60 W Carbon Filament Lamp (1905)	4.0
60 W Coiled Coil Tungsten Filament Lamp (1968)	14.7
1000 W General Service Lamp (1961)	23.0
No.1 Photoflood Lamp (1961)	34.6
400 W Mercury Vapor Lamp (1968)	*57.5
40 W Fluorescent Lamp, Cool White (1968)	*80.0
96-Inch High Output Fluorescent Lamp (1968)	*82.0
400 W Metal Halide Lamp	*85.0
1000 W Metal Halide Lamp	*100.0
400 W High Pressure Sodium Lamp	*125.0
1000 W High Pressure Sodium Lamp	*130.0
180 W Low Pressure Sodium Lamp	*183.0
Note: * Values are for lamps only and do not include ballast losses	

Table #95 - Efficacy of Lamps

Lighting Terms (cont'd)

Footcandle: This is an Imperial measurement unit that quantifies the amount of illumination present on a surface. One footcandle is equal to 10.764 lux. Table #96 provides representative levels of illumination expressed in both footcandles and in lux.

Glare: Glare is an annoying and visually impairing encroachment of light into a person's plane of vision. Glare is the worst offender to visual comfort. Glare may be classified as being either direct, reflected, or veiling. Lighting systems should be designed such that little or no glare will be present.

Representative Illumination Levels		
	Footcandle	Lux
Starlight	0.0002	0.002
Moonlight	0.02	0.2
Street Lighting	0.6 - 1.9	6 - 20
Daylight - at North window	50 - 200	500 - 2000
Daylight - in shade (Outdoors)	100 - 1000	1000 - 10,000
Daylight - Direct Sunlight	5000 - 10,000	50,000 - 100,000
Office Lighting	70 - 150	750 - 1600

Table #96 - Representative Illumination Levels

Lighting Terms (cont'd)

Illumination: A measure of the level of light present on a surface and expressed in either footcandles (imperial system) or in lux (metric system). The recommended level of illumination depends on the task being performed, the age of the occupant, the speed at which a task will be performed and the importance of the task.

Tables #97 and #98 list recommended lighting levels for the type of task that will be performed. The values listed in tables #97 and #98 may require adjusting depending on other factors such as age of occupant, speed and accuracy etc.

Recommended Lighting Levels by Task			
Type of Visual Work	**Lighting Level**		**Comments**
	Footcandle	**Lux**	
Tasks occasionally performed	15	150	General Lighting
High contrast or large size	30	300	General Lighting
Medium contrast or small size	75	750	Task Lighting
Low contrast or very small size	150	1,500	Task Lighting
Low contrast or very small size - over a prolonged period	300	3,000	Combination of general and task lighting
Exacting visual tasks - very prolonged	750	7,500	Combination of general and task lighting

Table #97 - Recommended Lighting Levels by Task

Lighting Terms (cont'd)

Recommended Lighting Level by Task		
Building Area / Task	footcandle	lux
Auditoriums	15	150
Banks - teller stations	75	750
Barber shops	75	750
Bathrooms	30	300
Building entrances	5	50
Cashiers	30	300
Cleaning	15	150
Conference Rooms	30	300
Corridors	15	150
Dance Halls	7.5	75
Drafting - High contrast	75	750
Drafting - Low contrast	150	1,500
Elevators	15	150
Exhibition Halls	15	150
Floodlighting - Bright surroundings	30	300
Floodlighting - Dark surroundings	15	150

Building Area / Task	footcandle	lux
Hospitals - Examination Rooms	75	750
Hospitals - Operating Rooms	150	1,500
Kitchen	75	750
Laundry	30	300
Lobbies	30	300
Office - General	75	750
Parking Areas - Covered	5	50
Parking Areas - Open	2	20
Reading/Writing	75	750
Restaurant - Dining	7.5	75
Restaurant - Food Display	75	750
Stairways	15	150
Stores - Sales Areas	75	750
Street Lighting - Highways	1.5	15
Street Lighting - Roadways	0.5	5
Utility Rooms	30	300
Video Display Terminals	7.5	75

Table #98 - Recommended Lighting Levels by Area and Task

Lighting Terms (cont'd)

Infrared light: Nonvisible light used for radiant heating and commonly employed in heat lamps.

Light (visible): Radiant energy that excites the retina of the eye and creates the sensation of seeing.

Light loss factor: A numeric value representing the derating of a light source due to factors such as dirt, aging of the lamp, temperature, etc.

Luminaire (light fixture): A complete lighting unit which consists of a lamp(s), ballast (for discharge lamps), diffuser (louver or lens), socket(s), and internal wiring.

Luminaire efficiency: A ratio of light that is emitted by the luminaire to the total light produced by the lamps within the luminaire.

Lumen: A unit representing the quantity of light emitted by a lamp. A lumen may be thought of as a bundle or unit of light flux and may be likened to current in an electrical circuit. Representative values of lumens for various light sources are listed in table #99.

Representative Lumen Values for Lamps	
Light source (Lamp)	Lumen rating
100 W Incandescent	1750
100 W Quartz Halogen	2500
40 W Fluorescent - Cool White	3150
100 W Mercury Vapor	4300
100 W Metal Halide	8500
100 W High Pressure Sodium	9500
90 W Low Pressure Sodium	13,500

Table #99 - Lumen Values For Lamps

Lighting Terms

Lux: A metric measurement unit that quantifies the amount of illumination present on a surface. One lux is equal to 0.0929 footcandles. Table #96 provides representative levels of illumination expressed in both footcandles and in lux.

Reflectance: A percentage value of the light that is reflected from an illuminated object. Standard interior room reflectance values for calculation purposes are 80% ceiling, 50% wall and 20% floor.

Spacing ratios: A ratio number that represents the maximum distance between centerlines of adjacent luminaires and the mounting height of the luminaires.

Ultraviolet light (UV): Invisible light that causes phosphorous materials to fluoresce and produce visible light (UV-A), causes skin to redden (tanning UV-A and UV-B), and causes certain bacteria to be destroyed (irradiation UV-C).

Sources of Light

There are many processes available to convert electrical energy into visible light. Three common processes used today are incandescence, fluorescence and gas discharge. Incandescence involves heating a filament to a temperature that will result in visible light. Fluorescence involves creating of ultraviolet radiation and then exposing phosphor coated surfaces to the ultraviolet radiation resulting in visible light. Gas discharge lamps produce light by passing an electric current through a gas, which then causes the atoms in the gas to emit visible light.

Incandescent Lamps

An incandescent lamp produces light by heating to a high temperature a metallic filament made of tungsten (earlier filaments were made of carbon, osmium or tantalum), by means of an electric current.

Sources of Light (cont'd)

As shown in illustration #240, the tungsten filament is coiled and enclosed in a glass bulb that has been evacuated of air and replaced with a filling gas such as argon. The base shown is of the screw shell type and completes the connection to the electrical source.

One variation of the incandescent lamp is the tungsten quartz halogen lamp. This lamp is so named due to its construction which consists of a quartz glass outer envelope and the addition of a halogen element to the filling gas. When halogen is added to the filling gas, a chemical reaction occurs resulting in the evaporated tungsten being redeposited on the filament and thereby keeping the bulb wall free of tungsten blackening. The bulb of the tungsten halogen lamp is made of quartz glass to withstand the lamps high operating temperatures. Low voltage quartz halogen lamps operate at voltages of 6 or 12 V and incorporate stepdown transformers.

Incandescent lamps come in different shapes and sizes and are identified by their wattages and shape codes, as shown in illustration #241.

Table #100 lists some general characteristics of incandescent lamps.

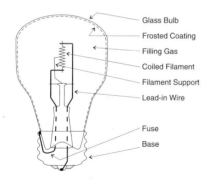

Glass Bulb
Frosted Coating
Filling Gas
Coiled Filament
Filament Support
Lead-in Wire
Fuse
Base

Illustration #240 - Construction of an Incandescent Lamp

Incandescent Lamps (cont'd)

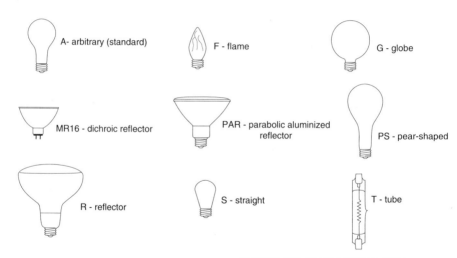

Illustration #241 - Incandescent Shape Codes

Incandescent Lamps (cont'd)

Characteristics of Incandescent Lamps	
Characteristic	**Description**
Lamp Watts	1 to 1500
Lamp Life (hours)	1,000 to 6,000 (typical 1000) Shortest life of all light sources (longer life lamps have a lower efficacy)
Luminous Efficacy	10 to 35 lumens per watt. Lowest efficacy of all light sources. Efficacy increases with lamp size
Lamp Lumen Depreciation Factor (LLD)	80 - 90 %
Color Temperature	2,500 to 3,000 K Warm color
Color Rendering Index (CRI)	97 Excellent CRI
Warm-up Time	Instant
Restrike Time	Instant
Lamp Cost	Lowest initial cost Highest operating cost
Main Applications	Residential, Retail display lighting, Galleries

Table #100 - Incandescent Lamp Characteristics

Fluorescent Lamps

A fluorescent lamp, as shown in illustration #242, operates by first heating the tungsten filament electrodes located at each end of the lamp; then the heated electrodes emit electric particles (electrons) which travel at high speed through the lamp and collide with electrons of the vaporized mercury atoms. Invisible ultraviolet light is produced as a result of these collisions. The ultraviolet light activates the phosphor and produces visible light. The color of the light produced depends on the type of phosphor used in the lamp.

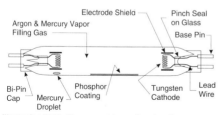

Illustration #242 - Fluorescent Lamp Construction

Electrode Shield
Pinch Seal on Glass
Argon & Mercury Vapor Filling Gas
Base Pin
Bi-Pin Cap
Mercury Droplet
Phosphor Coating
Tungsten Cathode
Lead Wire

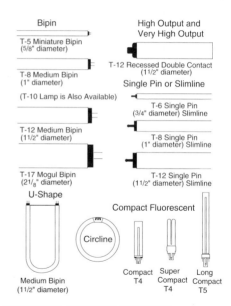

Bipin

T-5 Miniature Bipin
(5/8" diameter)

T-8 Medium Bipin
(1" diameter)

(T-10 Lamp is Also Available)

T-12 Medium Bipin
(1 1/2" diameter)

T-17 Mogul Bipin
(2 1/8" diameter)

High Output and Very High Output

T-12 Recessed Double Contact
(1 1/2" diameter)

Single Pin or Slimline

T-6 Single Pin
(3/4" diameter) Slimline

T-8 Single Pin
(1" diameter) Slimline

T-12 Single Pin
(1 1/2" diameter) Slimline

U-Shape

Medium Bipin
(1 1/2" diameter)

Circline

Compact Fluorescent

Compact T4

Super Compact T4

Long Compact T5

Illustration #243 - Fluorescent Lamp Examples

Fluorescent Lamps (cont'd)

All fluorescent lamps require a ballast to operate. The ballast is used to provide the lamps with their starting voltage and to limit the lamps operating current. There are three basic types of fluorescent lamps each with its own type of ballast; preheat lamps, instant start lamps, and rapid start lamps. The most common lamp is the rapid start, as shown in illustration #242. Rapid start lamps can be used for dimming and flashing applications and are available with either an electromagnetic (core and coil) ballast or the newer electronic ballast. Fluorescent lamps are available in various sizes and shapes, as shown in illustration #243.

Color codes are used on fluorescent lamps. Each manufacturer has a unique color code but certain generic codes are also used, as shown in table #101.

Fluorescent Lamp Generic Color Codes	
Color Code	**Description**
CW	Cool White
CWX	Cool White Deluxe
WW	Warm White
WWX	Warm White Deluxe
D	Daylight
N	Natural
C50	Colortone 50 or Chroma 50
C75	Colortone 75 or Chroma 75
R	Red
G	Green
B	Blue

Table #101 - Fluorescent Lamp Generic Color Codes

Fluorescent lamps are available in various lengths ranging from compact fluorescents to 8 foot tubular. Compact fluorescent lamps are now available in shapes that will allow for retrofitting of existing incandescent lamps.

Fluorescent Lamps (cont'd)

Fluorescent lamps are very sensitive to ambient temperature variations and light output will diminish if the temperature is either above or below their rated optimum value, which is usually 25°C (77°F). For low or high temperature operation, HO (high output) or VHO (very high output) lamps are needed.

Fluorescent ballasts are sound rated "A" (quietest) to "F" (loudest) and are either high or low power factor rated. Electronic ballasts are quieter and more efficient than electromagnetic types and are easier to dim. Illustration #244 shows a typical wiring arrangement for a two-lamp rapid-start circuit using an electromagnetic ballast.

Table #102 lists some general characteristics of fluorescent lamps.

Characteristics of Fluorescent Lamps	
Characteristic	**Description**
Lamp Watts	5 to 215 W
Ballast Watts (losses)	13 W for a typical F40T12 lamp
Lamp Life (hours)	20,000 for typical F40T12 lamp. 20 times the life of a typical incandescent
Luminous Efficacy	40 to 80 lumens per watt
Lamp Lumen Depreciation Factor (LLD)	70 - 90 %
Color Temperature	2,700 t0 7,500 K Wide range of color temperatures
Color Rendering Index (CRI)	62 to 94. Higher CRI means lower efficacy
Warm-up Time	Instant. Sensitive to extremes in temperature
Restrike Time	Immediate
Lamp Cost	Low. Reduced energy lamps more expensive
Main Applications	Offices, Commercial

Table #102 - Characteristics of Fluorescent Lamps

Fluorescent Lamps (cont'd)

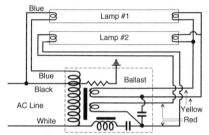

Illustration #244 - Rapid Start Lighting Circuit

High Intensity Discharge Lamps

Light is produced by passing current through a metal vapor at high pressure. These vapors consist of vaporized metal or metallic salts, such as mercury and sodium. A sealed arc tube, as shown in illustration #245, is used to hold electrodes and enclose the metallic vapours and starting gas.

Mercury vapor, metal halide and high pressure sodium are all classified as being high intensity discharge lamps. Each lamp works on the same operating principle but their vaporized metals differ, thereby producing a different type of light.

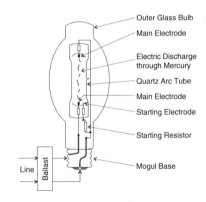

Illustration #245 - Mercury Vapor Lamp

High Intensity Discharge Lamps (cont'd)

Mercury vapor lamps: A ballast is required to start the lamp and to control the operating current. When the ballast is energized, an electrical field is established between the main electrode and the nearby starting electrode. Electrons are emitted, developing local glow and ionizing the argon starting gas.

A diffuse argon arc between main electrodes is created and the heat gradually vaporizes the mercury which joins the arcstream. The mercury vapor now takes over and changes the color from an argon blue glow to the bluish-green of mercury. Light intensity increases gradually and reaches full output after 5 - 7 minutes. Restrike time is between 3 to 10 minutes because the lamp must first cool down sufficiently to reduce the vapour pressure to a point where the arc will restrike. Mercury vapor lamps also have the following characteristics:

a. Available in sizes from 40 to 1,000 W.

b. Available in self-ballasted style from 160 to 1250 W.

c. Average lamp life is 24,000+ hours.

d. Lamps are available in either clear or phosphor coated.

e. Color temperatures available between 4,000 and 7,000 K.

f. CRI values are between 22 and 44.

g. Efficacies range from 20 to 60 lumens per watt.

h. Applications are limited to areas where a high color temperature is needed (exterior flood lighting, etc.).

i. Available in shape codes of A, BT, E, PAR, R, and T.

j. The symbol H is used to designate a mercury vapor lamp (eg. H33 400/DX).

Note: Mercury vapor lamps are rarely used for new lighting systems. Their poor efficacy and CRI make them undesirable and are now being replaced by metal halide and high pressure sodium lamps.

High Intensity Discharge Lamps (cont'd)

Metal Halide lamps: A ballast, as shown in illustration #246, is required to start the lamp and to regulate the operating current. When the ballast is first energized, the lamp will behave like a mercury vapor lamp. However as the arc tube temperature rises, the metallic salts also vaporize and begin to provide most of the visible light radiation. Several color steps may now be observed. The halides in the arc tube expand the color band and introduce a wider color spectrum than in the mercury vapor lamp. Warm-up time is about 5 minutes and restrike time is about 15 minutes.

Note: Metal Halide lamps have specified operating positions. Horizontal burning lamps have the arc tube bowed upward to follow the natural curve of the arc stream, and vertical burning lamps have an expanded arc tube to accommodate the convection currents generated by the arc stream.

Mount lamps only in the position designated by the manufacturer.

Metal Halide lamps also have the following characteristics:

a. Available in sizes from 32 to 1,500 W.

b. Lamp life ranges from 10,000 to 20,000 hours (the lower the wattage the lower the lamp's life).

c. Lamps are available either in clear or color corrected coated.

d. CRI varies from 65 to 80+.

e. Color temperature ranges from 3200 to 4200 K.

f. Efficacies range from 50 to 100 lumens per watt.

g. Typical applications include sports stadium lighting, supermarket lighting, sign lighting, outdoor building lighting, showroom lighting and industrial lighting.

h. Metal halide lamps are effective replacements for mercury vapor lamps.

High Intensity Discharge Lamps (cont'd)

i. Metal halide lamps are available in the following shape codes: BT, T.

j. Metal halide lamps are prefixed with the letter "M".

Note: Metal Halide lamps may NOT always operate on mercury vapor ballasts.

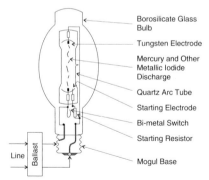

Borosilicate Glass Bulb

Tungsten Electrode

Mercury and Other Metallic Iodide Discharge

Quartz Arc Tube

Starting Electrode

Bi-metal Switch

Starting Resistor

Mogul Base

Line

Ballast

Illustration #246 - Metal Halide Lamp

High Pressure Sodium (HPS) lamps: A ballast, as shown in illustration #247, is required to start the lamp and to regulate the operating current. The operation of the lamp is similar to that of the mercury vapor and metal halide lamps except that the lamp has no starting electrode. Therefore, the ballast contains a special starting device (ignitor) that provides a high voltage spike which establishes the xenon gas arc between the main electrodes. Mercury and sodium then rapidly vaporize and change the cool xenon and mercury colors to a warm golden white tone (warm-up time is 4 to 6 minutes). The starting pulse provided by the ballast ignitor allows the lamp to restrike within one minute versus the 4 to 15 minutes that it takes for a mercury vapor or a metal halide lamp.

High Pressure Sodium lamps also have the following characteristics:

a. Lamps are available from 35 to 1000 W.

b. Lamp life is 24,000+ hours.

High Intensity Discharge Lamps (cont'd)

c. HPS lamps are available in clear or coated for color correction.

d. Color temperatures range from 1900 to 2700 K.

e. CRI values range from 20 to 80+.

f. Efficacies range from 50 to 130 lumens per watt and vary with lamp size (lowest efficacies occur with lower wattage lamps).

g. Available in the following shape codes:T, B, PAR, BT, E.

h. HPS lamps may be mounted in any operating position.

i. HPS lamps that are designed with an internal ignitor may be directly retrofitted into an existing mercury vapor or metal halide luminaire.

j. The letter S is commonly used as a prefix for the lamps.

k. Typical applications for HPS lamps include: street lighting, high and low bay industrial lighting and commercial space lighting where warmer colors are needed.

Note: Cycling or inoperative high pressure sodium lamps should be replaced promptly to avoid stress and possible damage to the ignitors and ballasts. An ignitor continues to pulse until a lamp is lighted and, in time, will fail unless the inoperative lamp is replaced or the power is turned off.

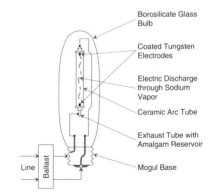

Illustration #247 - High Pressure Sodium Lamp

High Intensity Discharge Lamps (cont'd)

Low Pressure Sodium (LPS): These lamps are more closely related to fluorescent lamps than HID, since it is a low pressure, low intensity discharge source, and has a linear lamp shape. The LPS lamp consists of either a U-shaped arc tube, as shown in illustration #248A, or a linear lamp arc tube, as shown in illustration #248B. This arc tube is constructed of sodium-resistant lime borate glass. The arc tube is placed inside a glass envelope that has an internal coating of heat-retaining indium oxide (this coating contributes to the very high efficacy of the lamp). Inside the arc tube is a mixture of neon and argon gases, which are used to start the lamp, together with pure sodium metal. When the lamp first starts, virtually all the sodium is condensed, and the current is carried by the neon-argon starting gas producing a red glow.

As the lamp warms up sodium is vaporized and the discharge begins to exhibit the characteristic yellow color of a LPS lamp. Warm-up time is 7 to 9 minutes, and the restrike time is immediate.

Low Pressure Sodium lamps also have the following characteristics:

a. Available in lamp sizes ranging from 18 to 180 W.

b. Lamp life is 12,000 (18 W lamp) and 18,000 (all others) hours.

c. The lamp is monochromatic (one color only) yellow.

d. The CRI is extremely poor. Everything is either yellow or muddy in appearance.

e. Color temperature is 1700 K.

f. Lamp efficacies range from 100 to 180 depending on lamp size (lower wattage - lower efficacy).

g. Typical applications for LPS are security lighting, area floodlighting where color is not important and some road lighting.

High Intensity Discharge Lamps (cont'd)

h. Unlike all other lamps, LPS lamps do not depreciate in light output with age.

Note: LPS lamps must be operated within their specified operating positions. The lower wattage lamps (18 to 55 W) may be operated vertically with the base up, and higher wattage lamps (90 to 180 W) may be operated horizontally. Any significant variation from the specified operating position may result in shortened lamp life.

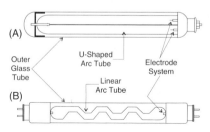

Illustration #248 - Low Pressure Sodium Lamp

Maintenance and Troubleshooting

In order to achieve optimum performance of a lighting system, regular maintenance needs to be performed. Regularly scheduled checks of all lamps and their condition is important in stemming the failure of other parts of the lighting system. Regular cleaning of both the lamps and the luminaire will improve light output. Group relamping will save time and result in higher levels of light output. On occasion, lighting equipment may fail or experience abnormal behavior, resulting in less than optimum performance. The following troubleshooting tables list symptoms, possible causes and suggested remedies for incandescent, fluorescent, HID and LPS lighting systems.

Maintenance and Troubleshooting - Incandescent Lamps		
Symptom	**Possible Causes**	**Suggested Remedies**
short life	-overvoltage	-change voltage rating of lamp to that of source
	-shock or vibration	-replace with rough service lamp
	-improper burning position	-replace existing lamp with lamp designed for mounting position
loose base	-excessive operating temperatures	-check if lamp wattage is suitable for luminaire
bulb blistering or bulging	-excessive operating temperatures	-check if lamp wattage is suitable for luminaire
	-slow leak	-replace lamp

Table #103 - Troubleshooting Incandescent Lamps

Maintenance and Troubleshooting - Fluorescent Lamps		
Symptom	**Possible Causes**	**Suggested Remedies**
normal lamp failure	-emissive material on tungsten electrodes has been depleted	-replace lamp
immediate failure of lamp	-defective lamp, possibly an air leak	-replace lamp
lamps operate at unequal brilliancy	-variation in ambient temperature	-protect lamps from operating in drafts or cold temperature. Provide ventilation where heat is the problem.

Table #104 - Troubleshooting Fluorescent Lamps

Symptom	Possible Causes	Suggested Remedies
blinking on and off	-normal end of life	-promptly remove old lamp
	-loose circuit contact (at socket)	-seat lamp securely, tighten all connectors
	-cold drafts or winds hitting tube	-enclose or protect lamp
short life	-too frequent starting of lamps	-life expectancy is based on 3 hours per start. If lamps are turned off and on more frequently, life is shorter
	-high or low voltage	-check power supply
	-one lamp of a two-lamp circuit burned out and second lamp operating dimly. Dim lamp will fail early	-immediately replace burned-out lamp
	-severe end blackening caused by open electrode heater circuit	-check for poor contact between lamp pins and socket or loose socket connection and break in line
 Brownish Rings at one or both ends about 2 inches from base	-happens to some lamps	-none needed
 Dense black spots about 1 inch from base about a half inch wide extending half way around tube	-appears early in life and does not increase in size or appears near end of life	-none needed as lamp performance is okay

Table #104 - Troubleshooting Fluorescent Lamps (cont'd)

Symptom	Possible Causes	Suggested Remedies
Dark streaks lengthwise along tube	-globules of mercury condensed on lower part of tube	-the lower half of tube is cooler than upper half; by rotating lamp 180° these globules should evaporate
Dark streaks or reduced light output	-cold draft hitting tube or low temperature operation	-enclose or protect the lamp
Dense end blackening at one or both ends within 1 inch of the base and extending 2-3 inches or blackening early in life	-normal end of life -poor contact between lamp pins and socket	-replace lamp -check for proper socket spacing or poor socket construction not providing proper wiping of pins when lamp is installed
Blackening generally within 1 inch of the end. Blackening early in life	-Mercury deposit -rapid evaporation of emission material on electrodes due to loose contact -wrong lamp type	-should evaporate as lamp is operated -ensure lamp holders are rigidly mounted and the lamp is secure -replace with correct lamp

Table #104 - Troubleshooting Fluorescent Lamps (cont'd)

Symptom	Possible Causes	Suggested Remedies
no starting effort or slow starting	-defective lamp	-replace lamp
	-open circuit	-check connections and socket seating
	-low circuit voltage	-check power supply
	-dirt accumulation on lamps which overcomes effect of silicone coating and may thereby cause unreliable starting at high humidity	-wash lamps with water containing a mild detergent and then rinse
	-temperature below 10°C(50°F)	-use low temperature ballast
	-luminaire not properly grounded	-install bonding conductor
swirling, spiraling, snaking or fluttering of arc stream	-occurs in new lamps	-should stabilize in normal operation or after the lamp has been turned on and off a few times
	-defective lamp if condition persists	-replace lamp
decreased light output	-cold drafts hitting lamp	-enclose or protect lamp
	-starting and regulating of lamp is defective	-replace ballast
	-dirt or dust on lamp, luminaire or room surfaces	-clean surfaces

Table #104 - Troubleshooting Fluorescent Lamps (cont'd)

Maintenance and Troubleshooting - Mercury Vapor and Metal Halide Lamps		
Symptom	**Possible Causes**	**Suggested Remedies**
lamp will not start	-end of life	-replace lamp
	-poor socket contact	-check lamp contact
	-photocell defective	-replace photocell
	-low supply voltage	-check power source
	-low socket voltage	-check ballast tap and connections
	-incorrect or failed ballast	-replace ballast
	-abnormally low ambient temperature	-install ballast rated for low temp.
lamp has abnormally short life	-overwattage or excessive lamp current and may be evidenced by burnt seals, melted connector ribbons, bulged arc tube or discolored outer bulb due to failed or improper ballast	-replace ballast or lamp
	-ballast incorrect for supply voltage	-install properly rated ballast
	-shorted capacitors	-replace capacitors

Table #105 - Troubleshooting Mercury Vapor and Metal Halide Lamps

Symptom	Possible Causes	Suggested Remedies
lamp flickering or cycling on/off	-incorrect ballast	-replace ballast or lamp
	-low-ballast open-circuit voltage	-check power supply, check ballast
	-supply voltage dip	-place lighting on separate circuit
	-poor electrical connections	-check wiring and socket connections
lamp emits a faint glow	-insufficient open circuit voltage	-check voltage and ballast
	-hard starting lamp	-replace lamp
light output low	-incorrect ballast	-match ballast with lamp
	-incorrect tap or low supply voltage	-correct tap or raise supply voltage to meet ballast input requirement
	-dirty lamp or luminaire	-clean lamp and luminaire regularly
	-normal depreciation	-replace lamp or group relamp
	-failed capacitor	-replace capacitor

Table #105 - Troubleshooting Mercury Vapor and Metal Halide Lamps (cont'd)

Maintenance and Troubleshooting - High Pressure Sodium Lamps		
Symptom	**Possible Causes**	**Suggested Remedies**
lamp will not start	-end of life -poor socket contact -photocell defective -low supply voltage -low socket voltage -incorrect or failed ballast -ballast ignitor defective	-replace lamp -check lamp contact -replace photocell -check power source -check ballast tap and connections -replace ballast -replace ignitor
lamp has abnormally short life	-overwattage or excessive lamp current and may be evidenced by burnt seals, melted connector ribbons, bulged arc tube or discolored outer bulb due to failed or improper ballast -ballast incorrect for supply voltage -shorted capacitors	-replace ballast or lamp -install properly rated ballast -replace capacitors
lamp flicker, cycling on/off	-incorrect ballast -normal end of life -high lamp operating voltage -low ballast open circuit voltage -supply voltage dip -poor electrical connections	-replace ballast or lamp -replace lamp -replace lamp -check power supply, check ballast and replace if necessary -place lighting on separate circuit -check wiring and socket connections
lamp emits a faint glow	-insufficient open circuit voltage -hard starting lamp	-check voltage and ballast -replace lamp

Table #106 - Troubleshooting High Pressure Sodium Lamps

Symptom	Possible Causes	Suggested Remedies
light output low	-incorrect ballast -normal depreciation -failed capacitor -incorrect tap or low supply voltage supply -dirty lamp or luminaire	-match ballast with lamp -replace lamp or group relamp -replace capacitor -correct tap or raise voltage to meet ballast input requirement -clean lamp and luminaire regularly
lamp breakage, ring-offs at neck	-glass strain in bulb -excessive vibration -socket or luminaire scratches bulb or lamp is overtightened	-replace lamp -correct cause of vibration -replace lamp or socket and do not overtighten lamp when inserting into the socket
lamp color variation	-variations in luminaire cleanliness	-clean luminaires and if necessary replace refractor or cover glass

Table #106 - Troubleshooting High Pressure Sodium Lamps (cont'd)

Maintenance and Troubleshooting - Low Pressure Sodium Lamps		
Symptom	Possible Causes	Suggested Remedies
lamp will not start	-end of life -poor socket contact -photocell defective -low supply voltage -low socket voltage -incorrect or failed ballast	-replace lamp -check lamp contact -replace photocell -check power source -check ballast tap and connections -replace ballast

Table #107 - Troubleshooting Low Pressure Sodium Lamps

Symptom	Possible Causes	Suggested Remedies
lamp has abnormally short life	-refer to table #106	-
lamp flickering or cycling on/off	-incorrect ballast -low-ballast open-circuit voltage -supply voltage dip -poor electrical connections	-replace ballast or lamp -check power supply, check ballast and replace if necessary -place lighting on separate circuit -check wiring and socket connections
lamp emits a faint glow	-insufficient open circuit voltage -hard starting lamp	-check voltage and ballast -replace lamp
lamp emits pink glow	-normal end of life	-replace lamp
light output low	-incorrect ballast -incorrect tap or low supply voltage -dirty lamp or luminaire	-match ballast with lamp -correct tap or raise supply voltage to meet ballast input requirement -clean lamp and luminaire regularly
lamp breakage, ring-offs at neck	-glass strain in bulb -excessive vibration	-replace lamp -correct cause of vibration

Table #107 - Troubleshooting Low Pressure Sodium Lamps (cont'd)

SECTION ELEVEN

MEASURING INSTRUMENTS

Basic Instruments

Voltage, current, resistance and power measurements are routinely made in electrical circuits. Electrical instruments are used to measure and monitor these circuit values. The most common field instruments used by electricians to test and/or troubleshoot electrical circuits are voltmeters, ammeters, ohmmeters, megohmmeters, and occasionally wattmeters.

Often three instruments are combined into one single instrument known as a VOM, volt-ohm-milliammeter, as shown in illustration #249. These multimeters are designed to allow the user to choose the type of electrical unit (current, voltage, resistance) to be measured by means of a selector switch. Multimeters (VOM) are available in analog (needle movement) or in digital (DMM-digital multimeter with discreet numbers appearing on a screen) form.

In addition to these instruments, there are hundreds of other measurement devices used to measure, monitor, analyze, and interpret the behavior of electrical circuits.

VOM (Volt-Ohm Milliammeter)

DMM (Digital Multimeter)

Illustration #249 - Multimeters

Voltmeter

To measure voltage, a voltmeter (either AC or DC) is connected across an electrical component (observe polarity marks when measuring DC). This connection is known as a parallel connection. The voltmeter as shown in illustration #250A consists of a meter movement connected in series with some resistors. The meter movement is usually rated for 50 mV with the rest of the circuit voltage being dropped across the series resistors.

It is very important that the correct scale be chosen in order that the meter movement not be subjected to voltages higher than it is capable of safely handling.

The series resistors shown in illustration #250A are used to limit the current through the meter movement to 1 mA when the rated voltage is applied across the voltmeter. A lesser voltage would reduce the current in the circuit and the deflection on the meter would be less.

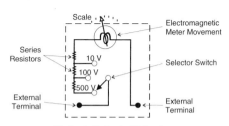

(A) Basic Parts of a Voltmeter

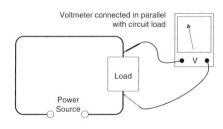

(B) Voltmeter Connection

Illustration #250 - Multirange Voltmeter

Voltmeter (cont'd)

A voltmeter's sensitivity or accuracy is rated in ohms-per-volt. The higher the ohms-per-volt rating the greater the sensitivity or accuracy. The accuracy is calculated by dividing full scale current into 1 volt; eg. 1 V / 1 mA = 1000 ohms-per-volt.

Note: Voltmeters are always connected across (parallel to) circuit components (illustration #250B), never in series. A voltmeter is a high impedance device, and if it is connected in series with a circuit component the behavior of the electrical circuit will change. Use very high impedance voltmeters to measure voltage in high impedance electronic circuits.

Ammeter

The ammeter as shown in illustration #251A consists of a basic meter movement and shunting resistors. The meter movement for the ammeter is the same as that of the voltmeter (1 mA), and therefore almost all of the measured current must be shunted.

Multirange ammeters have shunting resistors of different values depending on the size of the measured current. An ammeter is always connected *into* the electrical circuit (series connection) as shown in illustration #251B if of the two lead type (either AC or DC); or placed around the circuit conductor as shown in illustration #251C if of the clip on type (also known as a tong-test ammeter).

When inserting a two lead ammeter into an electrical circuit first open the circuit at the point where the current measurement is desired and then insert the ammeter (observe polarity marks when measuring DC values).

Note: Ammeters are always connected in series with electrical circuit components and must never be connected across the electrical circuit. An ammeter is a very low impedance device, and if connected across an electrical circuit large fault currents will flow.

Ammeter (cont'd)

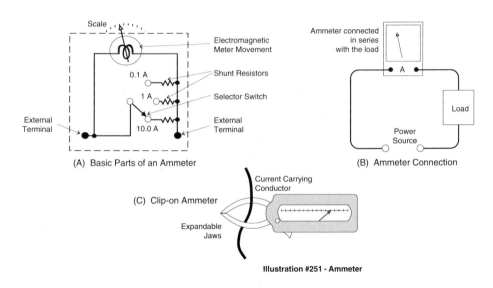

(A) Basic Parts of an Ammeter

Scale

Electromagnetic Meter Movement

0.1 A

1 A

10.0 A

Shunt Resistors

Selector Switch

External Terminal

External Terminal

(B) Ammeter Connection

Ammeter connected in series with the load

A

Load

Power Source

(C) Clip-on Ammeter

Current Carrying Conductor

Expandable Jaws

Illustration #251 - Ammeter

Ohmmeter

An ohmmeter, shown in illustration #252, consists of a battery and a variable resistor connected in series with a basic meter movement. The variable resistor is used to calibrate the meter to obtain a full scale deflection (zero ohms showing on the meter) when the two terminal leads are shorted together. Calibration is done to compensate for changes in the internal battery voltage due to aging (only needs to be done on analog type ohmmeters).

Note: Each time an analog ohmmeter's scale is changed the meter must be recalibrated.

When the ohmmeter leads are open, the pointer is located at the far left side of the scale indicating infinite (∞) resistance (open circuit) (illustration #253A). When the two leads are touched together, the pointer deflects completely to the right indicating zero resistance (short circuit) (illustration #253B).

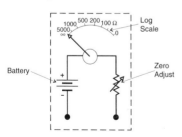

Illustration #252 - Basic Ohmmeter

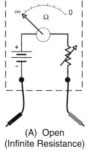 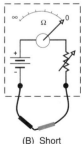

(A) Open
(Infinite Resistance)

(B) Short
(Zero Resistance)

Illustration #253 - Ohmmeter Indications

Ohmmeter (cont'd)

Note: An ohmmeter has its own power supply (battery) and therefore must never be connected into a circuit that is energized or being acted upon by another voltage source. If an ohmmeter is connected into an energized circuit, there is a high risk of damaging the meter and/or the electrical circuit being measured.

The following steps are recommended when using an ohmmeter to measure resistance:

1. Disconnect the circuit from the voltage.
2. Isolate the circuit component being measured to prevent other circuit devices from altering the meter reading.
3. Select the proper range switch setting. It is best to obtain readings that will cause analog type ohmmeters to deflect on the right half of the scale (ohmmeters are based on a log scale). For digital meters, set the switch at the lowest setting that is greater than the value expected.
4. For analog meters, touch the meter leads together, and use the ohms adjust dial to set the pointer at zero ohms.
5. Connect the meter leads across the circuit component being measured. Polarity is not important when measuring resistors; however, for some electronic circuit components, such as diodes, polarity does matter. Ensure that fingers do not touch the test lead as body resistance will alter the meter reading.

Note: For precision resistance measurements in the range of 1 to 100,000 Ω, a Wheatstone Bridge is often used (see Section One, page 30).

Megohmmeter

A megohmmeter, popularly known as a "megger", is an instrument that is used for measuring very high resistance values. The term megohmmeter is derived from the fact that the device measures resistance values in the megohm (1 megohm = 1,000,000 Ω) range.

Megohmmeter (cont'd)

The primary function of the megohmmeter is to test insulation resistance of power transmission systems, electrical machinery (motors, generators), transformers and cables. A basic megohmmeter consists of a hand-driven generator and a direct-reading true ohmmeter as shown in illustration #254. The unknown resistance is connected between the terminals marked line and earth.

A hand crank is turned at a moderate speed (approx. 120 RPM), and a DC voltage is generated. The scale is calibrated so that the pointer directly indicates the value of the resistance being measured (all values are shown in megohms). The purpose of the guard ring (third terminal) is to eliminate surface current leakage across exposed conductors and/or ground. Illustration #255 shows a connection diagram for testing insulation resistance values for a cable with the third terminal (guard ring) being used.

Note: The guard terminal is at the same potential as the line terminal

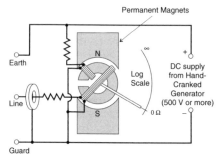

Illustration #254 - Megohmmeter Construction

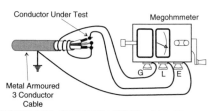

Illustration #255 - Cable Testing With Megohmmeter

Megohmmeter (cont'd)

Note: Megohmmeter models exist which produce different levels of test voltage (100, 250, 500, 1000, 1500, 2500, 5000 or 10,000 V). The megohmmeter may be hand cranked, motor driven or battery supplied to develop the intended DC voltage. Because the amount of power that the megohmmeter can produce is small, the test is ordinarily considered to be nondestructive, meaning permanant damage is not likely to be caused in the insulation system of the device being tested. However the level of output voltage is high enough to be a personnel safety hazard if incorrect testing procedures are used. Never connect a megohmmeter insulation tester to energized lines or equipment.

Wattmeter

Wattmeters measure the power of an electrical circuit as shown in illustration #256. The wattmeter is effective in both AC and DC circuits.

A wattmeter consists of two electromagnetic coils with one coil connected across (parallel) the electrical circuit being measured and the other coil being connected into (series) the electrical circuit. The interaction of the magnetic fields of these two coils will result in a net value of power ($P = V \times I$).

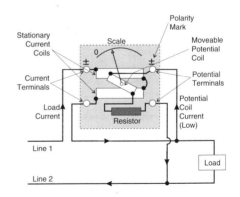

Illustration #256 Wattmeter Construction

Wattmeter (cont'd)

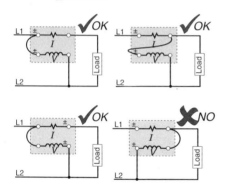

Illustration #257 Connections of Wattmeters

Wattmeters have polarity marks which must be observed in order that correct and meaningful readings are obtained. Illustration #257 shows both the correct and incorrect ways in which wattmeters may be connected.

Remember that the current should (normally) enter at the polarity mark and leave at the nonpolarity mark for both the voltage and current coils.

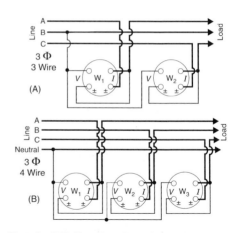

Illustration #258 - Three Phase Power Measurements

Wattmeter (cont'd)

Power in three phase circuits may be measured by using two or more wattmeters. Illustration #258 shows the proper connections for measuring power in both a three phase three wire system and a three phase four wire system. The power readings obtained by the connections shown in illustration #258A are correct for both balanced and unbalanced loads that are operating at power factors of 50% or better. The total power of the circuit is obtained by adding the values shown on the wattmeters.

Polyphase wattmeters are available that will replace the two or three individual units shown in illustration #258 with one unit. Energy measurement is accomplished by the use of watthour meters which make use of small motors in which the speed of rotation is proportional to the power.

Indicating dials are rotated by the action of the watthour meter, and a kWh value is displayed as shown in illustration #150, page 259 of Section Six.

Note: The potential (voltage) coil of the wattmeter must always be connected across (parallel) to the electrical circuit being measured, and the current coil must always be connected into the electrical circuit (series). The voltage measurement part of the wattmeter has a very high resistance, and the current measurement part of the wattmeter has a very low resistance. If improperly connected, the behavior of the electrical circuit may change, and/or very high fault currents may flow. The polarities of each coil must be observed and proper connections made in order to obtain readings that are both meaningful and correct.

SECTION TWELVE
SYMBOLS AND DRAWINGS

Electrical Symbols

Standardized symbols of equipment and wiring should always be used in electrical drawings. By using standardized symbols, the electrical drawing may be easily interpreted by others. Symbols provide quick and easy access to critical information concerning the electrical circuit or system. Electrical symbols may be thought of as graphic codes that when assembled onto a drawing convey information in a condensed form.

ANSI, IEEE and CSA electrical symbol standards have been developed to form a common graphic code language for drawings. There are however variations to these standards, and no mandated (code required) symbols have been developed.

The following symbols are commonly used in electrical drawings in North America:
table #108 schematic symbols,
table #109 single line symbols and
table #110 floor plan wiring diagram symbols.

A more comprehensive listing of symbols may be found in the following standards:

ANSI Y32.2-1975,
IEEE 315-1975,
CSA Z99.3-1979.

SCHEMATIC SYMBOLS

SYMBOL	DESCRIPTION	SYMBOL	DESCRIPTION	SYMBOL	DESCRIPTION
	Normally Open Push Button		Two Position Selector Switch		Normally Open Temperature Switch
	Normally Closed Push Button		Three Position Selector Switch		Normally Closed Temperature Switch
	Normally Open, Normally Closed Push Button		Normally Open Liquid Level Switch		Plugging Speed Switch
	Single Pole Single Throw Switch		Normally Closed Liquid Level Switch		
	Single Pole Double Throw Switch		Normally Open Pressure Switch		Anti - Plugging Speed Switch
	Double Pole Double Throw Switch		Normally Closed Pressure Switch		
			Normally Open Contact		Normally Open Limit Switch
	Double Pole Single Throw Switch		Normally Closed Contact		Normally Closed Limit Switch

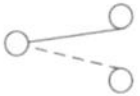

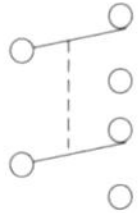

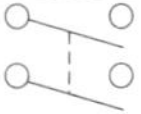

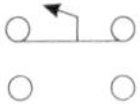

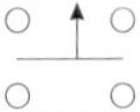

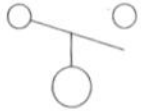

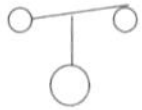

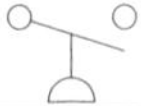

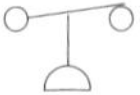

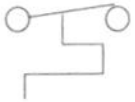

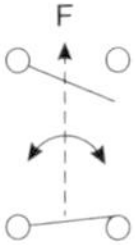

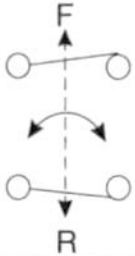

Table #108A - Schematic Symbols

SCHEMATIC SYMBOLS

SYMBOL	DESCRIPTION	SYMBOL	DESCRIPTION	SYMBOL	DESCRIPTION
	Normally Open Time Delayed on Energization Contact		Normally Open Foot Switch	(AM)	Ammeter
	Normally Closed Time Delayed on Energization Contact		Normally Closed Foot Switch	(VM)	Voltmeter
	Normally Open Time Delayed on De-energization Contact		Normally Open Flow Switch		Transformer
			Normally Closed Flow Switch		Capacitor
	Normally Closed Time Delayed on De-energization Contact		Relay Coil		Inductor
			Relay Current Coil		Fuse
			Lamp		Overload Heater
			Bell		Battery
			Horn or Siren		Ground
			Buzzer		Control Wire
		RES	Resistor		Power Wire
					Mechanical Connection
					Connected Wiring
					Not Connected Wiring

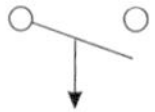

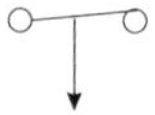

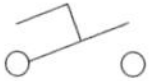

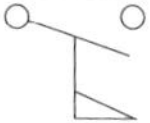

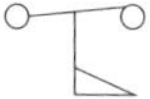

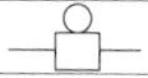

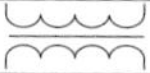

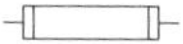

Table #108B - Schematic Symbols

SINGLE LINE DIAGRAM SYMBOLS

SYMBOL	DESCRIPTION	SYMBOL	DESCRIPTION	SYMBOL	DESCRIPTION
	Disconnectable or Withdrawable Device		Power Transformer	SV	Solenoid Valve (mechanical operation should be noted)
1200A 52	Power Circuit Breaker		Potential Transformer	87	Protective Relay. Number is Device Number
600A	Circuit Breaker		Current Transformer	G	Lamp- G = Green R = Red W = White etc.
600A	Disconnect Switch		Delta Connection	★	Meter * V-Voltmeter A-Ammeter W-Wattmeter t° -Temperature
600A	Fused Disconnect Switch		Open Delta Connection		
600A	Fused Isolating Switch	Y	Wye Connection	○ Red ○ Start ○ Stop	Push Button Station Start - Stop With Red Indicating Lamp
	Magnetic Motor Contactor Complete With Overload Heaters	• —●—‖‖	Lightning Arrester		
≪≪◻≫≫	Fuse in Pullout Type of Disconnect	20	Squirrel-cage Motor. Number Denotes Horsepower	⏚	Ground

Table #109 - Single Line Diagram Symbols

SYMBOLS FOR FLOOR PLAN WIRING DIAGRAMS

	SYMBOL	DESCRIPTION	SYMBOL	DESCRIPTION	SYMBOL	DESCRIPTION
RECEPTACLES, OUTLETS & SWITCHES	—Ⓒ	Clock Outlet	▲	Floor Telephone Outlet	$\$_4$	Four Way Switch
	⊖	Duplex Receptacle	△	Special Purpose Single Outlet as Noted	$\$_K$	Key Operated Switch
	⊖ GF	Duplex Receptacle Protected by GFI	Ⓣ	Thermostat	$\$_P$	Switch & Pilot Light
	⊖	Split - Switched Duplex Receptacle	Ⓟ	Photo Cell	$\$_D$	Dimmer Switch
	⊖ WP	Weatherproof Receptacle	▲	Telephone Outlet Wall Mounted	Ⓢₐ	Low Voltage Switch a = Switch Identification
	⊖	Split - Duplex Receptacle	Ⓗ	Humidistat	$\$$ a b	Two Gang Switch a, b = Switch Identification
	⊖ R	Range Receptacle	TV	T.V. Outlet	$\$$ a $\$$ b	Same as Above
	⊖ CD	Clothes Dryer Receptacle	$\$$	Single Pole Switch	●	Pushbutton
	⊡	Floor Single Receptacle	$\$_3$	Three Way Switch	☐○	Bell

Table #110A - Floor Plan Wiring Diagram Symbols

SYMBOLS FOR FLOOR PLAN WIRING DIAGRAMS						
	SYMBOL	DESCRIPTION	SYMBOL	DESCRIPTION	SYMBOL	DESCRIPTION
WIRING, RACEWAYS, STARTERS & MISC.	CH	Chime	———	Wiring in Ceiling, Wall	⬚	Fused Disconnect
	T	Transformer	- - - - -	Wiring Exposed	⊠	Combination Magnetic Motor Starter
	F	Fire Alarm Station	— — — -	Wiring in Floor	⊠	Magnetic Motor Starter
	F ◯	Fire Alarm Bell	A1,3 ⫽➔	Number of Strokes = Number of Conductors Arrow = Home Run A1, 3 - Panel A Circuits 1 and 3	M	Meter Base
	◑	Smoke Alarm			$^M	Manual Motor Starter
	◑	Smoke Detector	—◯	Raceway Up	Ⓜ ★	Motor ★ = H.P. Rating
	◕	Heat Detector	—●	Raceway Down		
	●	Heat Detector - High Temperature	⌐	Raceway Capped	Note: Some Symbols Shown in Tables 110A, B & C are Non - Standard	
	ELR	End of Line Resistor	⬚	Disconnect		

Table #110B - Floor Plan Wiring Diagram Symbols

SYMBOLS FOR FLOOR PLAN WIRING DIAGRAMS

	SYMBOL	SYMBOL	DESCRIPTION	SYMBOL	SYMBOL	DESCRIPTION
	Ceiling	Wall		Surface	Flush	
OUTLETS, LUMINAIRES & PANELBOARDS	E	—E	Emergency Outlet	◢◢◢	◢◢	Power Panel
	J	—J	Junction Box	■■■	■■	Lighting Panel
	X	—X	Exit Light	◺	◺	Telephone Panel
	F	—F	Fan Outlet Fractional H.P. Motor	▭	▭	Special Purpose Panel as Noted
	b ⌐A3 (1)	b —A3 —(1)	Luminaire Outlet 1 = Luminaire ID Number b = Switch A3 = Panel A Circuit 3			
	▭R	—▭R	Fluorescent Luminaire R - Recessed			
	├─────┤	├─┴───┤	Fluorescent Strip Luminaire			

Table #110C - Floor Plan Wiring Diagram Symbols

Electrical Drawings

Many types of diagrams may be needed to explain, describe, and detail the operation and construction of an electrical system. There are four common types of electrical drawings:

- Schematics
- Equipment wiring diagrams
- Single line or three line diagrams
- Floor plan wiring diagrams

Schematics

Schematic drawings are used to understand the operation of an electrical circuit and for trouble shooting.

A schematic diagram reveals the electrical components, their connections and how they are related to each other electrically. A schematic should be referred to when trying to understand how a circuit works.

Schematic diagrams are read like a book. Start reading the schematic at the upper left-hand corner, and then proceed from left to right and from top to bottom. Complex schematics will have each rung or line numbered and cross referenced to other rungs or lines.

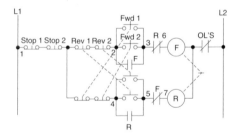

Illustration #259 - Schematic Drawing

Schematics (cont'd)

When drawing schematics, the following rules should be followed:

1. All lines are to be drawn either vertically or horizontally and should proceed from left to right.
2. Relay or contactor coils should be identified with numbers or letters.
3. Contacts should be identified as to the relay or the contactor responsible for their operation.
4. Wherever possible, all terminals should be numbered.
5. Schematics should be drawn so that a minimum number of lines cross over each other.
6. Junction points and crossovers should be clearly distinguishable so that no confusion arises.
7. Mechanical interlocks are identified by a dotted line.
8. Contacts, switches or operating devices are to be shown in their deenergized state or as they would be when not acted upon by any force.
9. Use standard symbols, and always accompany the drawing with a symbol legend.

Equipment Wiring Diagrams

Wiring diagrams show the physical connections of electrical devices in an electrical circuit. There are three common types of wiring diagrams:

1. Point to point diagrams show the terminal connection location and routing path of every wire used in the system. Each line represents a wire which is shown starting at one point and ending at another, as shown in illustration #260.

Equipment Wiring Diagrams (cont'd)

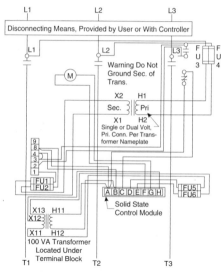

Illustration #260 - Point to Point Wiring (Sample)

2. Baseline diagrams are drawings that feed all wires into one central line called a baseline, as shown in illustration #261.

These drawings present the wiring in a well organized easy to follow manner but with the disadvantage that they do not show components as they are physically located. These drawings are most often used for maintenance and assembly manuals.

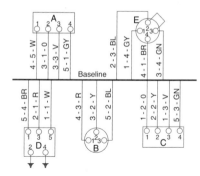

Illustration #261 - Baseline Wiring Diagram (Sample)

Equipment Wiring Diagrams (cont'd)

3. Highway diagrams combine groups of wires running along similar paths into bundles called highways, as shown in illustration #262. Components are shown in their relative positions, and all conductors and circuit devices are labelled for easy tracing.

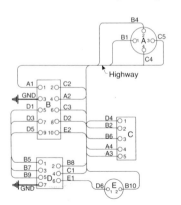

Illustration #262 - Highway Wiring Diagram

Single Line and Three Line Diagrams

Single line diagrams are diagrams that graphically represent an electrical system by single lines. These single lines can represent two, three or more actual conductors and are simply intended to show interconnections between system components rather than the actual wiring connections. A single line diagram helps to trace the power flow of the electrical system and to obtain an overview of the major parts of the electrical system. The single line diagram shown in illustration #263A may also be drawn in two or three line form, as shown in illustration #263B, depending on the number of phases present (see table #109 for single line diagram symbols).

Floor Plan Wiring Diagrams

A floor plan wiring diagram is a drawing that defines and shows the location of various rooms and areas in a building or site.

Floor Plan Wiring Diagrams (cont'd)

The floor plan is drawn to scale, and the electrical wiring is then placed on to it as shown in illustration #264. Standard floor plan symbols should be used and each set of floor plans must be accompanied by a symbol legend.

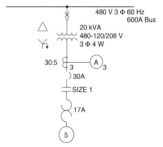

Illustration #263A - Single Line Diagram

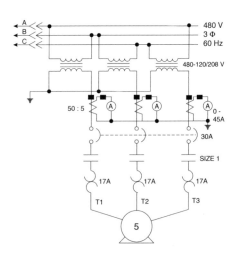

Illustration #263B - Three Line Diagram

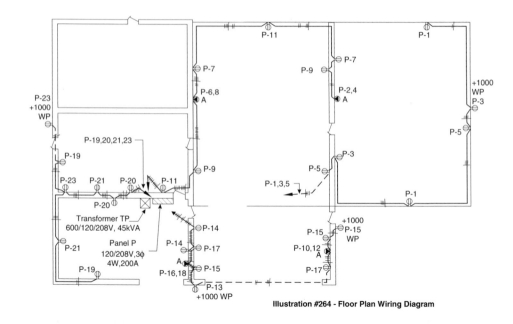

Illustration #264 - Floor Plan Wiring Diagram

APPENDIX
EQUATIONS, ABBREVIATIONS AND FORMULAS

Ohms Law:

a. $I = V/R$

b. $V = IR$

c. $R = V/I$

Resistance:

a. $R_T = R_1 + R_2 + R_3 + ... + R_N$
(total resistance in series)

b. $1/R_T = 1/R_1 + 1/R_2 + 1/R_3 + ... + 1/R_N$
(total resistance in parallel)

c. $R_T = R_1 R_2 /(R_1 + R_2)$ (total resistance of two resistors in parallel)

d. $R_T = R/n$ (total resistance of n numbers of the same resistors in parallel)

e. $R_x = R_2 R_3 /R_1$ (Wheatstone bridge)

Inductance:

a. $L_T = L_1 + L_2 + L_3 + ... + L_N$
(total inductance in series)

b. $1/L_T = 1/L_1 + 1/L_2 + 1/L_3 + ... + 1/L_N$
(total inductance in parallel)

c. $L = N^2 \mu \, ^A/$ (inductance in terms of physical parameters)

d. $\tau = L/R$ (time constant)

e. $W = 0.5 L I^2$ (energy stored by an inductor)

f. $X_L = 2 \pi f L$ (inductive reactance)

Capacitance:

a. $C_T = C_1 + C_2 + C_3 + ... + C_N$
(capacitors in parallel)

b. $1/C_T = 1/C_1 + 1/C_2 + 1/C_3 + .. + 1/C_N$
(capacitors in series)

c. $C = 8.85 \times 10^{-12} \, A k / d$ (capacitance in terms of physical parameters)

d. $\tau = RC$ (time constant)

e. $W = 0.5 \, C V^2$
(energy stored by a capacitor)

f. $X_C = \dfrac{1}{2\pi f C}$ (capacitive reactance)

Resistance, Inductance and Capacitance in Series AC Circuits:

a. $X = X_L - X_C$ (net reactance)

b. $Z = \sqrt{R^2 + X^2}$ (impedance)

c. $S = \sqrt{P^2 + Q^2}$ (apparent power)

d. PF = P/S (power factor)

e. $V_T = \sqrt{V_R^2 + V_X^2}$ (total voltage)

f. $f_R = 1/(2\pi\sqrt{LC})$ (resonant frequency)

Resistance, Inductance and Capacitance in Parallel AC Circuits:

a. $I_T = \sqrt{I_R^2 + I_X^2}$ (total current)

b. $Z_T = V_T/I_T$ (impedance)

c. $S = \sqrt{P^2 + Q^2}$ (apparent power)

d. PF = P/S (power factor)

e. $f_R = 1/(2\pi\sqrt{LC}) \times \sqrt{(X_L/R)^2/((X_L/R)^2+1)}$ (resonant frequency)

Power:

a. $P = V_R I_R$, $P = I_R^2 R$, $P = V_R^2/R$ (true power)

b. $Q = V_X I_X$, $Q = I_X^2 X$, $Q = V_X^2/X$ (reactive power)

c. $S = V_T I_T$, $S = I_T^2 Z$, $S = V_T^2/Z$ (apparent power)

d. $P_T = P_1 + P_2 + P_3 + ... + P_N$ (true power in series or parallel)

e. $Q_T = Q_1 + Q_2 + Q_3 + ... + Q_N$ (reactive power in series or parallel)

f. $S_T = S_1 + S_2 + S_3 + ... + S_N$ (apparent power in series or parallel)

AC:

a. $V_{RMS} = 0.707\ V_{PEAK}$ (effective voltage)
b. $V_{avg.} = 0.637\ V_{PEAK}$ (average voltage)
c. $E_{inst} = E_{max}\ \sin \theta$
(instantaneous voltage values)

Three-Phase:

a. $Vp = V_L / \sqrt{3}$
(wye-connected phase voltage)
b. $Vp = V_L$ (delta-connected phase voltage)
c. $Ip = I_L$ (wye-connected phase current)
d. $Ip = I_L / \sqrt{3}$
(delta-connected phase current)
e. $P = V_L I_L \sqrt{3} \times PF$
(true power for delta and wye)
f. $Q = V_L I_L \sqrt{3} \sin \theta$
(reactive power for delta and wye)
g. $S = V_L I_L \sqrt{3}$
(apparent power for delta and wye)

General:

a. $V_X = V_T (R_X / R_T)$ (voltage divider principle)
b. $I_X = I_T (R_T / R_X)$ (current divider principle)

Abbreviations

Abbreviations for Standards and Standard Agencies	
ANSI	American National Standards Institute
CEC	Canadian Electrical Code
CSA	Canadian Standards Association
EEMAC	Electrical and Electronic Manufacturing Association of Canada
FMA	Factory Mutual Approved
IEC	International Electro-Technical Commission
IEEE	Institute of Electrical and Electronic Engineers
IES	Illuminating Engineering Society
ISO	International Organization for Standardization
NEC	National Electrical Code
NEMA	National Electrical Manufacturing Association
NFPA	National Fire Protection Association
UL	Underwriters' Laboratories
ULC	Underwriters' Laboratories of Canada

Table #A1 - Standards Abbreviations

Conversion Factors

Conversion Factors			
From	**To**	**Factor**	**Reciprocal**
radians	electrical degrees	0.0174	57.47
radians/sec	rev/min	9.5493	0.10472
newtons	pounds	0.22481	4.4482
newton-metre	lb-ft	0.73755	1.3558
mm	inches	0.0394	25.4
metre	feet	3.2808	0.3048
mile	feet	5280	0.0001894
km	mile	0.6214	1.6093
litre	gal (US)	0.2642	3.785
litre	gal (Cdn)	0.2202	4.541
lux	footcandle	0.0929	10.764
kWh	joule	3,600,000	0.227×10^{-6}
hp	watt	746	0.00134

Table #A2 - Conversion Factors

Electrical Quantities and their Units and Symbols

Quantity	Symbol	Unit	Symbol
admittance	Y	siemens	S
angular velocity	ω	radian/sec	rad/sec
capacitance	C	farad	F
capacitive reactance	Xc	ohm	Ω
electric charge	Q	coulomb	C
electrical conductance	G	siemens	S
electric current	I,i	ampere	A
electric potential	V,v,E,e	volt	V
energy, work	W	joule	J
force	F	newton	N
frequency	f	hertz	Hz
illuminance	E	lux	lx
impedance	Z	ohm	Ω

Table # A3 - Quantities, Units, Symbols

Electrical Quantities and their Units and Symbols			
Quantity	Symbol	Unit	Symbol
inductance	L	henry	H
inductive reactance	X_L	ohm	Ω
luminous flux	Φ	lumen	lm
luminous intensity (candlepower)	I	candela	cd
power	P	watt	W
power, apparent	S	volt ampere	V•A
power, reactive	Q	reactive volt ampere	var
resistivity	ρ	ohm•metre or ohms per cmil-foot	Ω•m
resistance	R	ohm	Ω
susceptance	B	siemens	S
torque	T	newton metre	N•m
voltage	V,v	volt	V

Table #A4 - Quantities, Units, Symbols

Greek Symbols and their Electrical Designation			
Lower case letter	Upper case letter	Name	Electrical Designation
α	A	alpha	angle, coefficient
β	B	beta	angle, coefficient
γ	Γ	gamma	angle
δ	Δ	delta	angle, three-phase
ε	E	epsilon	dielectric constant, permittivity
η	H	eta	efficiency, hysteresis
θ	Θ	theta	angle
μ	M	mu	permeability
π	Π	pi	3.141592654
ρ	P	rho	resistivity
σ	Σ	sigma	sign of summation
τ	T	tau	time constant
ϕ	Φ	phi	magnetic flux, angle
ω	Ω	omega	angular velocity, resistance in ohms

Table #A5 - Greek Symbols